The
Quantum
World

The Quantum World

Michel Le Bellac

University of Nice, France

Foreword by

Alain Aspect

 sciences

 World Scientific

Published by

World Scientific Publishing Co. Pte. Ltd.

5 Toh Tuck Link, Singapore 596224

USA office: 27 Warren Street, Suite 401-402, Hackensack, NJ 07601

UK office: 57 Shelton Street, Covent Garden, London WC2H 9HE

Library of Congress Cataloging-in-Publication Data
Le Bellac, Michel.
    [Monde quantique. English]
    The quantum world / Michel le Bellac (University of Nice, France).
      pages cm
    Includes index.
    ISBN 978-9814579506 (hardcover : alk. paper) --
    ISBN 978-9814522427 (softcover : alk. paper)
    1. Quantum theory. I. Title.
    QC174.12.L42813 2013
    530.12--dc23
                            2013021877

British Library Cataloguing-in-Publication Data
A catalogue record for this book is available from the British Library.

Typeset by Stallion Press
Email: enquiries@stallionpress.com

Printed in Singapore by Fulsland Offset Printing (S) Pte Ltd

Preface

The object of this book is twofold. First I would like to present the basic principles of quantum mechanics and especially its foundational one, the superposition principle. Second, I would like to give some idea of the practical usefulness of this theory, by describing a few devices of everyday life: lasers, transistors, light emitting diodes, or atomic clocks used in GPS. These now-familiar objects could be built only thanks to our understanding of quantum physics. Finally, I shall also introduce the reader to some cutting edge research in quantum physics, for example, that devoted to ultra-cold atoms and to Bose–Einstein condensates.

The last thirty years witnessed an impressive progress in our understanding of the basic principles. One prominent feature of modern quantum mechanics is entanglement, which lies at the heart of the violation of Bell's inequalities, a result which compels us to revise in depth our ideas on relativistic space-time. Entanglement also sets the stage for quantum computers which could outperform classical ones in some specific, but very important problems, and will be the subject of Chapter 8. However, we should not underestimate the huge difficulties which must be overcome before we are able to build a really efficient quantum computer. Entanglement is also an essential feature in the interpretation of quantum mechanics. Indeed, while there is no controversy on the practical use of quantum mechanics, what it

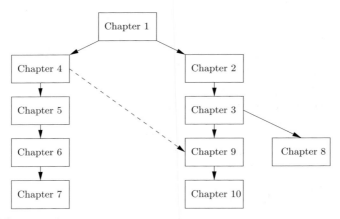

Figure 1. Logical flow chart for the different chapters. The left hand column corresponds to a reading more oriented toward applications, and the right hand one to a reading more oriented toward problems of principle.

really means is still hotly debated. I shall examine in some detail the issue of decoherence, which is a consequence of entanglement, and has strong implications for the famous measurement problem.

On the practical side, I clearly had to make choices, since quantum physics has an almost infinite number of applications. After having explained the basic features of atomic structure and of atom–light interaction, I shall sketch the basic design of a laser and describe the fundamental properties of laser light. Then I shall move on to low temperature physics and explain why cold atoms allow us to exert minute control over many devices and, as an example, I shall show how the accuracy of atomic clocks may be improved by a factor of 100 thanks to atomic fountains. A whole chapter is devoted to semiconductors, and the goal there is to understand the functioning of light emitting diodes (LEDs) and laser diodes, which are so important in everyday life for reading CDs and DVDs and for optical fiber communications. Relativistic quantum field theory, elementary particle physics and gravity will be briefly addressed in the following chapter.

The reader will find some equations in the main text, but they are purely algebraic and involve additions and multiplications only. High school math, with some vague notions of an exponential or of a sine function, is the only prerequisite to cope with the main text. More advanced notions, at the level of a first year undergraduate course in mathematics, are needed for some of the boxes and for the appendices, where some familiarity with complex numbers and vector spaces is useful.

The reader may follow two different approaches, according to his or her taste. The reader more interested in fundamental principles may follow the right hand part of the flow chart, or the left hand part if he or she is more interested in the practical aspects.

Acknowledgments First of all, I would like to thank Alain Aspect who agreed to write up a foreword for this book. I am especially grateful to Sébastien Tanzilli and Mathieu Le Bellac, who read in detail the whole manuscript and made a lot of useful comments. I also wish to thank Thierry Grandou, Franck Laloë, Michèle Leduc, Emmanuel Mennini and Borge Vinter for their remarks, and George Batrouni and Dan Ostrowsky for their help with the translation. Mathieu Le Bellac drew Figure 1.2 and Michel Laget supplied me with Figure 1.4

Nice, March 2013

Foreword

Writing a popular book is always a difficult task. When this book addresses quantum physics, a world in which our bearings and everyday experiences do not allow us to build intuitive representations, it becomes a major challenge. This is the challenge Michel Le Bellac decided to take up. He must be congratulated for that, because quantum physics witnessed major upheavals during the past decades, and it is essential that a general audience can find books which allow them to understand what is involved when they hear of entanglement, decoherence, and Bose–Einstein condensates or of quantum cryptography and quantum computers.

Quantum physics was born at the beginning of the twentieth century, with the foundational works of Planck, Einstein, Bohr and de Broglie, soon followed by those of Heisenberg, Schrödinger and Dirac who collectively devised a coherent framework and a mathematical formalism still in use today. Quantum physics allows us to describe the entirety of microscopic phenomena which lie beyond the scope of classical electrodynamics, the synthesis of the two great theories completed during the nineteenth century, mechanics on the one hand and electromagnetism on the other. Indeed, the greatest of physicists such as Lorentz failed in their attempts to understand the stability of matter, known at that time to be made of positive and negative charges which, by attracting each other, should have provoked the collapse of matter on itself. Rutherford's atomic model, a kind of solar system where electrons orbit the nucleus much as planets orbit the Sun, was not immune to this objection, because an electric charge which is forced to change its direction emits radiation: this is the principle behind synchrotron radiation. The outcome is that this charge loses energy and finally falls on the nucleus, exactly as a satellite, slowed down by friction forces in the upper layers of the atmosphere, finally falls to Earth. Elaborating on the discrete properties of radiation introduced by Einstein and Planck, Niels Bohr made the radical assumption that electron orbits could, in a similar way, take only specific discrete values determined from Planck's constant, whose value was deduced from measurements performed on blackbody radiation. The interpretation of orbit quantization was given by de Broglie and Schrödinger in terms of matter waves which, like vibrations of a guitar string, can only take specific frequencies. Wave mechanics was born. At the same time, Heisenberg developed an entirely different formalism, called matrix mechanics, which led to the same values of the wavelengths of radiation emitted by the hydrogen atom, in agreement with the spectroscopic measurements. In a dazzling synthesis, Schrödinger showed the equivalence of the two approaches and Dirac introduced a superbly elegant formalism, which we still use at present. The modern form of quantum mechanics was born. It had countless successes, as it finally allowed us to understand the structure of matter and its stability, as well as its mechanical, electrical, thermal and optical properties. It also allowed us to account for breathtaking phenomena occurring at very low temperatures, such as helium superfluidity, that is, the complete absence of resistance of this liquid to flow, and superconductivity, the fact that some materials can carry an electrical current without any loss. Finally, just after World War II, quantum mechanics led to two discoveries, which were to radically transform our societies, the transistor and the laser. Without computers, whose performances originate in the integration of an immense number of transistors, and without optical fiber communications made possible by lasers, we would not have entered the era of information and communication. We can then speak without exaggeration of a *quantum revolution* in order to refer to this impressive collection of advances, first in our understanding of the structure and properties of matter, and then in novel technologies, from information superhighways to computers to magnetic resonance imaging (MRI).

This scientific and technological revolution of the first part of the twentieth century was accompanied by another one, just as drastic, in our representation of the world. With quantum mechanics, we had to admit that particles behave sometimes as waves. Conversely, light, whose main properties are those of an electromagnetic wave, sometimes behaves as a flow of particles, or photons. This *wave–particle duality* lies at the heart of most of the quantum phenomena which were known in the 1960s, so that Richard Feynman did not hesitate to write on this subject, in his famous "Lectures on Physics" (Feynman *et al.* [1965]):

> ". . .a phenomenon which is impossible, *absolutely* impossible, to explain in any classical manner, and which has in it the heart of quantum mechanics. In reality, it contains the *only* mystery."

It turned out that the 1960s were going to see the emergence of a *new quantum revolution*, based on the realization of the importance of another quantum concept, that of *entanglement*, introduced on the one hand by Einstein and his colleagues Podolsky and Rosen, and on the other hand by Schrödinger. Entanglement is a property of quantum particles being able to form a whole system which cannot be correctly described by giving only all the properties of its individual constituents. The whole is more than the collection of the parts, and this holds true even if the constituents are located far away from each other in space-time, in such a way that relativity forbids any communication between them, whatever the interaction. This is translated into the existence of correlations much stronger than those allowed by classical physics, as is shown by the experimentally confirmed violation of Bell's inequalities. We had to wait until the discovery of these inequalities, in 1964, in order to realize the importance of entanglement, and it is to Bell's inequalities that Feynman refers to when he finally writes in 1982:

> "I've entertained myself always by squeezing the difficulty of quantum mechanics into a smaller and smaller place, so as to get more and more worried about this particular item. It seems to be almost ridiculous that you can squeeze it to a numerical question that one thing is bigger than another. But there you are — it is bigger than any logical argument can produce. . ."

Beyond the conceptual revolution implied by the realization of the drastically novel character of entanglement, a scientific revolution was going to follow immediately, whose beginnings were explicitly found in Feynman's article which we just quoted, namely, the possibility of devising *quantum computers*, whose computing power would be exponentially larger — in the literal meaning of the word — than that of any classical computer. At about the same time, from the 1960s on, physicists became able to isolate, control and observe *individual quantum objects*, be it an electron, a photon, an ion, an atom or a molecule. The collection of all these conceptual and experimental breakthroughs allowed physicists to develop rapidly a novel field of research, *quantum information*, whose purpose is to use the most extraordinary resources of quantum physics, and in particular entanglement, for information processing and transmission. It is too early to know whether the remarkable progress already achieved will result in the huge technological revolution which would correspond to the implementation of a quantum computer, but the consequences of such a success would amply justify the expression "second quantum revolution".

Michel Le Bellac's book has the great merit of highlighting consequences from recent breakthroughs, and of presenting in a synthetic way the concepts which underlie the two quantum revolutions, as well as a number of remarkable phenomena which can only be understood within this conceptual framework. The reader will find a description of interference experiments performed with single photons, illustrating the "mysterious" wave–particle duality, as well as application of quantum physics to semi-conductor lasers which read CDs and DVDs. A whole chapter is devoted to entanglement, to the Bohr–Einstein debate and to the test of Bell's inequalities, another chapter to quantum cryptography, and another one to laser cooling of atoms and the famous atomic Bose-Einstein condensates. The state of the art of research on quantum computers is reported lucidly, without concealing the difficulties which must be overcome in order to implement a useful device. Finally, the reader will find in this book not the answer, which we still don't know as of today, but a clear formulation of difficult questions far from being solved, on the quantum-to-classical frontier and on the interpretation of quantum mechanics. I was lucky enough to work on some of these questions, and I can attest that Michel Le Bellac has, on each of these questions, been able to bring to the fore the essential points and to choose relevant examples without swamping the reader with an accumulation of phenomena. This demanding book requires sustained attention, but it is worth it as the subjects which are addressed are among the most exciting in present research. The reader will be amply rewarded for his or her efforts. He or she will be in a better position to follow the discoveries which will, without any doubt, be announced in the coming years.

Michel Le Bellac, whose initial background is elementary particle physics, has for many years endeavored to understand in detail research which goes far beyond his own specialty. He has already written textbooks at an advanced level. Here, he offers his vast culture to a general public interested in science and its recent discoveries. He should be thanked for this. I hope this book will be successful, which would attest the renewed interest aroused by the "quantum world".

Alain Aspect

Contents

1

An Inconvenient Principle

At the end of the XIXth century, physics was dominated by two main theories: classical (or Newtonian) mechanics and electromagnetism. To be entirely correct, we should add thermodynamics, which seemed to be grounded on different principles, but whose links with mechanics were progressively better understood thanks to the work of Maxwell and Boltzmann, among others. Classical mechanics, born with Galileo and Newton, claimed to explain the motion of lumps of matter under the action of forces. The paradigm for a lump of matter is a *particle*, or a *corpuscle*, which one can intuitively think of as a billiard ball of tiny dimensions, and which will be dubbed a *micro-billiard ball* in what follows. The second main component of XIXth century physics, electromagnetism, is a theory of the electric and magnetic fields and also of optics, thanks to the synthesis between electromagnetism and optics performed by Maxwell, who understood that light waves are nothing other than a particular case of electromagnetic waves. We had, on the one hand, a mechanical theory where matter exhibiting a discrete character (particles) was carried along well localized trajectories and, on the other hand, a wave theory describing continuous phenomena which did not involve transport of matter. The two theories addressed different domains, the only obvious link being the law giving the force on a charged particle submitted to an electromagnetic field, or Lorentz force. In 1905, Einstein put an end to this dichotomic wave/particle view and launched two revolutions of physics: special relativity and quantum physics. First, he showed that Newton's equations of motion must be modified when the particle velocities are not negligible with respect to that of light: this is the special relativity revolution, which introduces in mechanics a quantity characteristic of optics, the velocity of light. However, this is an aspect of the Einsteinian revolution which will not interest us directly, with the exception of Chapter 7. Then Einstein introduced the particle aspect of light: in modern language, he introduced the quantum properties of the electromagnetic field, epitomized by the concept of photon.

After briefly recalling the main properties of waves in classical physics, this chapter will lead us to the heart of the quantum world, elaborating on an example which is studied in some detail, the Mach–Zehnder interferometer. This apparatus is widely used today in physics laboratories, but we shall limit ourselves to a schematic description, at the level of what my experimental colleagues would call "a theorist's version of an interferometer".

1.1 Waves in classical physics

1.1.1   Surface waves in a pond

At least once in their lives, everybody has thrown a stone into a perfectly calm pond and observed a system of circular waves propagating on the surface from the point where the stone hit the water, which is the *source* of the system of circular waves. In order to characterize this wave phenomenon, we can place a floating cork at some distance from the source. The cork stays motionless before it is reached by the waves, then it begins jiggling up and down and finally stops moving when the pond is calm again. In order to obtain a well-controlled phenomenon, we shall replace the stone by a vertical rod moving vertically with a periodic motion which perturbs the water surface. Then, if the amplitude of oscillations is small enough, the trajectory of the cork is a small circle in a vertical plane whose center is fixed: on average, the cork does not progress with the wave. Wave dynamics is characterized by a *wave amplitude*: in the case of the cork, we can choose this wave amplitude as being the maximum vertical deviation of the cork position with respect to the pond surface in calm water. The wave amplitude will be denoted by a, it will be taken as positive above the pond surface

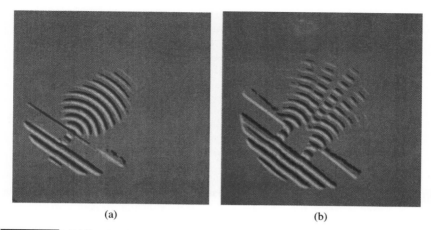

(a)   (b)

Figure 1.1. Diffraction and interference of surface waves in a pond. (a) Only one slit is open: one obtains a diffraction pattern. (b) The two slits are open: one obtains an interference pattern. One may observe the interference maxima (the "fingers") where the oscillation is strongest, and between the "fingers", regions where a cork would remain motionless. Reproduced from Hey and Walters [2003].

($a > 0$) and negative below it ($a < 0$). Because of the periodic character of the excitation, that is, the periodic motion of the rod, the vertical motion of the cork is also periodic: it reaches its maximum height at times separated by an interval T, the *period* of the oscillations.

Let us complicate slightly the experiment by setting a vertical wall between the point where the surface is perturbed by the oscillating rod and the point where the wave is observed thanks to the cork. The wall contains two slits so that the wave can propagate toward the cork, and the slits can be open or closed. If only one of the slits is open, we observe a *diffraction* pattern, that of the open slit (Figure 1.1a for experiment and 1.2a for a numerical simulation). Let us now open the two slits: the remarkable result is that the observed pattern cannot be deduced in a straightforward way from the two diffraction patterns. For example, if the cork oscillates at some point of the surface when slit 1 is open and slit 2 closed, and if it also oscillates in the opposite case with slit 1 closed and slit 2 open, it may happen that it stays motionless when the two slits are open! This phenomenon, a particular case of *interference*, (Figures 1.1b and 1.2b), finds its origin in that the full vibration amplitude of the cork is obtained by adding the amplitudes from the two slits, following rules which we are going to make explicit. In order to

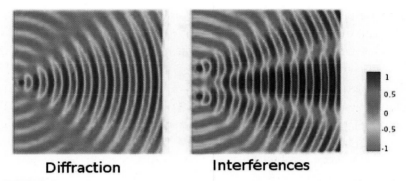

Diffraction   Interférences

Figure 1.2. Numerical simulation of diffraction on the left and of interference on the right at a fixed instant of time. A color code gives the height of the cork with respect to its position at equilibrium, that is in quiet water: this height lies between -1 (red) and $+1$ (blue): the oscillation amplitude is $+1$. One clearly sees on the right hand figure the two lines corresponding to destructive interference. Figure drawn by Mathieu Le Bellac.

understand these rules, let us concentrate on a fixed point of the surface. We call a_1 the wave amplitude at this point when slit 1 is open and slit 2 closed, and a_2 the amplitude corresponding to the opposite situation. Let us assume that at the point under consideration the two waves take their maximum value at the same time, that is, a crest from slit 1 coincides with a crest from slit 2: we then say that the two waves are "in phase". The full wave amplitude is the sum $a = a_1 + a_2$ of the two amplitudes, and we observe a *constructive interference*. The maximum height above the surface is given by the sum of the absolute values, $|a| = |a_1| + |a_2|$. If, at another point, a crest from slit 1 ($a_1 > 0$) coincides with a trough from slit 2 ($a_2 < 0$), we say that the two waves are out of phase. The absolute value of the full amplitude $a = a_1 + a_2$ is then $|a| = ||a_1| - |a_2||$. We observe a *destructive interference*, and, if $|a_1| = |a_2|$, the oscillation amplitude vanishes at that particular point. We clearly see in Figs. 1.1b and 1.2b the zones which correspond to destructive interferences. The previous discussion sheds light on a crucial property: if only one slit is open, we take into account a single oscillation amplitude and the only relevant information — besides the period of the oscillation — is the maximum height reached by the cork. However, when the two slits are open, and if we want to combine the two oscillation amplitudes, we must further have a measure of the time delay between the two waves. This time delay is proportional to an angle δ (see Section 1.5), which is the *phase shift* between the two amplitudes. Constructive interference corresponds to $\delta = 0$ and destructive interference to $\delta = \pi$. Intermediate cases will be treated later on, in Section 1.4.5 and Appendix A.1.4. The *intensity I* of the wave phenomenon is given by the square of the amplitude: $I = a^2$. The energy which is carried by a wave is proportional to its intensity, and thus to the amplitude *squared*. If two waves display a destructive interference, the intensity will be zero if $a_1 = -a_2$, and it will be four times the individual intensity in the case of constructive interference. These bizarre addition rules follow from the fact that we must add wave *amplitudes*, and not intensities.

1.1.2 Light waves

Surface waves are only a very particular case, and there exists a large variety of waves. Familiar examples are sound waves, which correspond to vibrations in gases, liquids or solids, and *light waves*, which correspond to oscillations of the electromagnetic field. In this latter case, one may identify the amplitude with that of the electric field. As in the case of surface waves, light waves are characterized by a period, an oscillation amplitude and an intensity I which is proportional to the electric (or magnetic) field *squared*.

Given that *electromagnetic waves* play a major role in what follows, let us briefly recall some of their properties. Electromagnetic waves propagate in a vacuum with the velocity of light, which is by definition $c = 299\,792\,458$ m/s $\simeq 3 \times 10^8$ m/s (see Appendix A.1.1 for the notation). As already mentioned, light waves are a particular case of electromagnetic waves, whose spectrum, which spreads from gamma rays to radio waves, is given in Figure 11.1 and in Appendix A.1.2. As in the case of surface waves, an important characteristic of an electromagnetic wave is its period T. Usually, one prefers to characterize a wave by its *frequency ν*, which is the inverse of the period: $\nu = 1/T$ and is measured in Hertz (symbol Hz). In an idealized situation, a wave displays a periodic pattern in space at a fixed instant of time, and the spatial periodicity is the *wavelength $\lambda = cT = c/\nu$*: it is the distance between two maxima of the amplitude. The wavelength of light waves lie between $\lambda = 0.40\,\mu m$ (violet) and $\lambda = 0.70\,\mu m$ (red), where $1\,\mu m = 1$ micrometer $= 10^{-6}$ m, formerly called a micron. Other very important wavelengths are, in the infrared, $1.31\,\mu$m and $1.55\,\mu$m, which correspond to absorption minima of electromagnetic waves by optical fibers, and which are, for this reason, almost universally used in optical fiber communications.

1.2 The Mach–Zehnder interferometer

Light waves, like surface waves in a pond, may display diffraction and interference patterns. A classic interference experiment is that of *Young slits*, which is the optics analogue of the experiment in Figure 1.1. This experiment is often used to introduce quantum physics, but I'd rather choose the *Mach–Zehnder interferometer* which, I believe, illustrates the basic principles even more directly. The interferometer is drawn schematically in Figure 1.3. A light beam, for example a laser beam, impinges from the left along a horizontal X direction upon a first beamsplitter BS_1, which makes a 45° angle with the X-direction. An ordinary window pane is an example

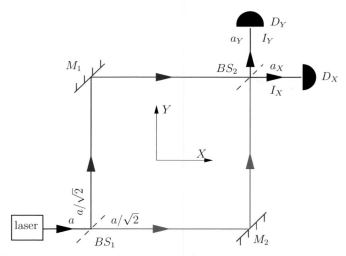

Figure 1.3.   The Mach–Zehnder interferometer. A laser beam travels horizontally on the lower left of the figure with a light intensity $I = a^2$. It is divided into two sub-beams of equal intensity (the *amplitudes* are thus divided by $\sqrt{2}$) by the beamsplitter BS_1, and each of the sub-beams is reflected either by mirror M_1 or mirror M_2 (green). The two sub-beams are recombined by the beamsplitter BS_2 and detected in D_X and D_Y, with intensities I_X and I_Y, $I = I_X + I_Y$.

of a *beamsplitter*: when looking at the pane from a suitable angle, one may see at the same time one's image (reflected light) and the external landscape (transmitted light). A non-absorbing beamsplitter reflects a fraction p of the incident light and transmits a fraction $1 - p$. The beamsplitter is *balanced* if $p = 1 - p = 1/2$. We shall assume that all the beamsplitters of our interferometer are balanced. Then beamsplitter BS_1 reflects 50% of the incident light, which follows a path parallel to the vertical Y direction (drawn in blue in Figure 1.3), and transmits 50% which travels along the horizontal, or X direction (drawn in red in Figure 1.3). The beams of the blue and red paths are finally recombined by a second beamsplitter BS_2 thanks to the intermediate mirrors M_1 and M_2. After BS_2, the intensity is divided into I_X in the horizontal beam and I_Y in the vertical beam. Since the mirrors, as well as the beamsplitters, are assumed to be perfectly non-absorbing, the total output intensity must be equal to the input one: $I = I_X + I_Y$.

As already mentioned, the intensity is proportional to the amplitude squared, and we shall write $I = a^2$, ignoring for simplicity any proportionality factor. The intensity of the two beams emerging from beamsplitter BS_1 is $I/2$, which means that the *amplitudes* are divided by $\sqrt{2}$. We shall assume for simplicity that the blue and red paths have exactly the same length. The light which arrives on detector D_X through the blue path is reflected by the first beamsplitter and transmitted by the second one, and the opposite holds for the red path. The paths followed by light are thus exactly identical (same length and same number of reflections and transmissions), which means that the amplitudes must be added for the path following the X direction after the second beamsplitter. The full intensity is received by detector D_X, so that detector D_Y receives no light at all! In others words, we have a constructive interference for detector D_X and a destructive one for D_Y: the amplitude $a_Y = 0$.

Box 1.1. Constructive and destructive interference.

After a transmission and a reflection, each amplitude is divided by $(\sqrt{2})^2 = 2$. The amplitude incident on D_X is given by $a_X = a/2 + a/2 = a$, and the light intensity received by D_X is $a^2 = I$. The destructive interference on D_Y follows from the fact that the two paths are not identical: the light from the blue path suffers two reflections, while that from the red path suffers two transmissions. One can show in optics that two reflections multiply the amplitude by a factor -1: $a_Y = a/2 - a/2 = 0$.

1.3 Photons

1.3.1 Introduction to the concept of photons

Until the beginning of the XXth century, physicists were convinced that the wave theory of light, grounded in the vibrations of the electromagnetic field, was in perfect agreement with experiment. The corpuscle theory advocated by Newton, who assumed that a light beam was composed of a stream of particles traveling in a straight line in a vacuum, had fallen into disrepute as it was unable to explain diffraction and interference phenomena which were perfectly accounted for by the wave theory of light used in our discussion of the Mach–Zehnder interferometer. It is precisely this construction that Einstein put upside down in 1905. Starting from an argument in statistical mechanics, Einstein concluded that the wave theory could not explain everything and that light *also* displayed a particle aspect. Furthermore, Einstein's theoretical argument was able to give an explanation for a mysterious threshold effect observed in the photo-electric effect: the emission of electrons from a metal illuminated by light occurs only if the wavelength is short enough, a phenomenon in complete contradiction with the standard wave theory which does not predict any effect of this kind. In 1915, Millikan, who was up to that point a staunch opponent of Einstein's theory, completed a series of very precise measurements of the photo-electric effect and he could only admit that Einstein was right! Actually, we know today the limitations of Einstein's theory: it is quite possible to explain the photo-electric effect without giving up the standard wave theory of light, provided the quantum aspect of matter is properly taken into account. However, Einstein's intuition was correct: in some experiments, light may appear to be formed of elementary "particles of light", which are called *photons*, a terminology coined by Lewis only in 1926. In fact, we had to wait until 1977, more than 70 years after Einstein's hypothesis, to devise optics experiments whose results could not be explained by the wave theory and which proved without any doubt the existence of photons. It must be observed, though, that already in the 1950s, particle and nuclear physicists had been able to prove the existence of high energy photons, the very energetic photons associated with γ-rays (Figure 11.1 and Appendix A.1.2).

The reason why photons are not part of our daily experience is that the number of photons which we usually encounter is huge, and we can only see the collective effects, not the individual ones. The particle aspect vanishes for a continuous and large flux of light. To be specific, let us take as an example a laser diode emitting red light with a power $P = 1$ mW = 1 milliwatt. According to Einstein, the energy E of each photon in the light beam emitted by the laser diode is related to its frequency ν by the so-called *Planck–Einstein relation*

$$\boxed{E = h\nu}, \tag{1.1}$$

where h is the *Planck constant*

$$h = 6.63 \times 10^{-34} \, \text{J} \times \text{s} . \tag{1.2}$$

The frequency corresponding to red light is $\nu \simeq 3.8 \times 10^{14}$ Hz, and the energy of an individual photon in the beam is then 2.5×10^{-19} J. The number of photons per second going through a surface perpendicular to the beam, that is, the *photon flux*, is $P/h\nu \simeq 4 \times 10^{15}$, a huge number even for such a low intensity.

In order to illustrate the particle aspect of light, we have chosen an example borrowed from astrophysics. Figure 1.4 contains photographs of stars taken with a CCD camera, which is an array of photon detectors, and registers the arrival of individual photons. In astrophysics, CCD cameras

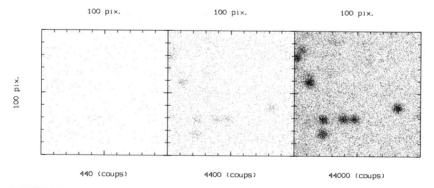

Figure 1.4. Photography of the same ensemble of stars with different exposures. In all cases the number of pixels is 10 000. The left hand photograph corresponds to the detection of 440 photons, the center one to 4400 and the right hand one to 44 000. Courtesy of Michel Laget.

replaced photographic plates in the early 1980s. On the left hand photo-graph, photons seem to arrive at random. When the number of photons increases, one observes that they arrive preferentially at points correspond-ing to the images of stars.

1.3.2   Photons and probabilities

One property of the particle aspect of light should be quite obvious, namely the probabilistic character of the impact of photons in Figure 1.4. As we just observed, in the left-most photograph of this figure, at first sight the pho-tons seem to hit the camera at random points, but they actually have a high probability of hitting a point where the image of a star is going to appear, and a low probability of hitting other points. This probabilistic aspect is fundamental, and it is worth giving a second illustration which uses a bal-anced beamsplitter (Figure 1.5a). Let us imagine that we put in the path of the laser beam of the previous subsection an attenuator which reduces the power of the beam by a factor 10^9, or down to 10^{-12} W. A simple calcu-lation using the numerical value of the velocity of light $c = 3 \times 10^8$ m/s

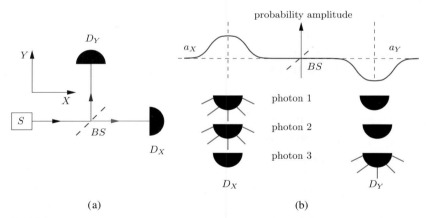

(a)   (b)

Figure 1.5.   (a) A source of single photons S sends photons into a balanced beamsplit-ter BS. The transmitted photons are detected by D_X and the reflected photons by D_Y: see Jacques *et al.* [2005]. (b) Schematic representation of probability amplitudes a_X and a_Y. For clarity, beamsplitter BS and detectors have been aligned on a horizontal axis, and the amplitude is given by the vertical axis. The full probability amplitude is the superposition of a_X and a_Y, and it is non-zero only in the vicinity of the detectors. The figure illustrates the case where it is assumed that the first and second photons trigger detector D_X and the third one triggers detector D_Y.

shows that the average distance between two photons is on the order of 100 m and it seems reasonable to assume that isolated, or single photons, are incident on the beamsplitter. In a hydrodynamical analogy, we go from a tap flowing continuously to a dripping tap. For reasons which are detailed later on, this is an intuitive, but not fully correct, way of producing single photons. Fortunately, sources which produce single photons efficiently are now available, and each photon can be tagged by the instant when it is produced. Let us now return to the beamsplitter of Figure 1.5a into which we send single photons. In order to simplify the argument, we assume that the detectors are 100% efficient: if a photon hits the detector, the detector is triggered in all cases, and there are no *dark counts*, that is, the detectors are never triggered in the absence of a photon — a dream for experimentalists! One then notices that a given photon triggers *only one* of the detectors, D_X or D_Y, and that this photon has a 50% probability of being transmitted toward detector D_X and a 50% probability of being reflected toward detector D_Y. It never happens that the two detectors are triggered simultaneously by a single photon, and this property can actually be used as a criterion to check that the photon source does produce single photons. Attenuated light does not have this property: it may happen that the two detectors are triggered simultaneously. Classical optics is recovered if we send a large number of photons: if, for example, we send 10 000 photons, about 5000 will be transmitted and 5000 reflected.

Box 1.2. Statistical fluctuations and polls.

More precisely, with a 70% probability, the number of transmitted photons will lie between 4900 and 5100. Indeed, statistical fluctuations in the number of transmitted (or reflected) photons are on the order of $\sqrt{10\,000} = 100$, that is, 1% in relative value. This is exactly the kind of fluctuations which are observed in a poll: with the standard sample of about 1000 persons, the error is on the order of 3% ($\sqrt{1000} \simeq 30$). A variation of 2% between two polls has little significance, being smaller that the statistical error. A so-called systematic error may have to be added to the statistical error, for example, if the sample is biased. The same phenomena are observed in an experiment whose results are governed by quantum physics. Since the number of registered events is necessarily limited, the probabilistic character of the theory implies that the accuracy of the experiment is limited by a statistical error which is all the smaller when the number of registered events is large. As in polls, one has also to take into account systematic errors due, for example, to imperfections in the apparatus.

The probabilities of reflection and transmission of a photon in the device drawn in Figure 1.5a are known, but the fate of an *individual* photon cannot be predicted: the only information which is available is its detection *probability* either by D_X or by D_Y. In quantum physics, probabilities are attached to *individual* quantum systems, while in classical physics, probabilities are attached to ensembles, a notion which will be defined shortly, and resorting to probabilities is a way of taking into account the complexity of phenomena of which we cannot (or do not want to) have detailed knowledge. For example, when tossing a coin, if we knew exactly the initial conditions (position of the coin, linear and angular velocities), if we took into account air resistance and had perfect knowledge of the landing ground, we could in theory predict the result, heads or tails. In practice, such a detailed information is not available, and we reason on *ensembles* of coins, that is, we toss a large number of identical coins, or we toss the same coin a large number of times. If the coins are not biased, probability theory correctly describes the results by assuming a 50% probability for heads and 50% for tails.

In classical probability theory, each individual in the ensemble is in a state where all its parameters are determined, even though they are unknown to us, but that is not the case in quantum physics where such parameters do not exist: the *probabilistic character of quantum physics cannot be attributed to our ignorance*. If, in the experiment of Figure 1.5a, two successive photons are prepared in a completely identical way, it may happen that the first photon triggers D_X and the second one D_Y, but we have no way of predicting the result in advance. In other words, quantum randomness is intrinsic, it is not linked to our lack of knowledge, and there is no ignorance interpretation of quantum probabilities. Therefore, we may use the device of Figure 1.5a for a sophisticated heads or tails game, with truly random results, and there exist commercial versions of this device which are used as generators of random numbers. Remember that computers generate "random numbers" from deterministic algorithms, and these numbers can only be pseudo-random. The random character of the results obtained with the device in Figure 1.5a is in complete contradiction with classical physics for which a full knowledge of the initial conditions and of the various forces allows us to foretell the future evolution, even if we know that this statement has only a limited value: actually, for systems which display chaotic behavior, two very close initial conditions may lead to two entirely different trajectories. However, we cannot exclude that, in the case of photons, we only have the *impression* of exactly identical preparation, but it might be that we have missed some properties: if we could access these properties, unknown to us for the time being and called accordingly *hidden variables*, then we could foretell the fate of each individual photon. We shall come back to this question in Chapters 3 and 10. Some physicists, among them de Broglie and Bohm, have proposed an explanation of this kind for the random character of quantum physics. However, we shall see later on why these hidden variables theories are not very popular among the majority of physicists.

1.3.3 Probability amplitudes and the superposition principle

Let us consider once more the device of Figure 1.5a but, instead of the single photon source, we use a laser beam sending into the beamsplitter an intensity I, which is split into a transmitted beam and a reflected beam. The intensities measured by the detectors D_X and D_Y are $I_X = I_Y = I/2$. To these intensities correspond wave amplitudes a_X and a_Y whose square gives the intensities. If we now return to the single photon source, the detection probabilities by D_X and D_Y are $\text{Prob}_X = 1/2$ and $\text{Prob}_Y = 1/2$. The probabilities for single photons are the analogue of the intensities for waves. It is then reasonable to ask the following question: is there for single photons an analogue of wave amplitudes? The answer to this question is positive: one can associate to the emission of a photon from a source, followed by its detection either by D_X or by D_Y, two corresponding numbers a_X or a_Y, whose square represents the probability of triggering one of the detectors, $\text{Prob}_X = a_X^2$ for D_X and $\text{Prob}_Y = a_Y^2$ for D_Y. Since their square represent probabilities, these numbers are called *probability amplitudes*. Figure 1.5b gives a schematic description of probability amplitudes: the amplitudes a_X and a_Y are non-zero only in the neighborhood of the detectors. Amplitude a_Y has been chosen to be negative to emphasize that probability amplitudes are not positive numbers in general. The mathematical truth is that probability amplitudes are complex numbers: see Box 1.3 and Appendix A.1.3. The reason why complex numbers cannot be avoided in quantum physics is given in Section 4.2. To keep the math simple, at least in the main text, we shall limit ourselves to real probability amplitudes, positive or negative.

The situation depicted schematically in Figure 1.5b illustrates a fundamental principle of quantum physics, the *linear superposition* (or *coherent
superposition*) of two quantum states: the photon is in a superposition of
two quantum states, the first one localized in the neighborhood of D_X and
the second one localized in the neighborhood of D_Y, but *only one of the
detectors is triggered*. This kind of superposition is not limited to photons:
one could easily perform the same experiment with neutrons, in which case
the beamsplitter would be a crystal, or with atoms, where the beamsplitter
would be a laser beam.

Box 1.3. Complex numbers.

If we want to be mathematically rigorous, we must be more precise: probability amplitudes are complex numbers, whose main properties are summarized in Appendix A1.3.
A probability amplitude a is written as the sum of a real part b and an imaginary part ic,
where b and c are real numbers, positive or negative, and $i^2 = -1$: $a = b + ic$.
The algebraic calculations with complex numbers follow the usual rules, taking into
account $i^2 = -1$. The *absolute value* $|a|$ of a complex number $a = b + ic$ is defined
as $|a| = \sqrt{b^2 + c^2}$, and the probability corresponding to amplitude a is $|a|^2$. In wave
theory, for example classical optics, complex numbers are just a convenient way to do
calculations, but they cannot be avoided in quantum physics.

One often explains the superposition of the two photon states which
we just encountered with the following intuitive picture: the photon (or
any other quantum particle such as a neutron or an atom) is at two different places at the same time. This is a useful picture, but it should not
be taken literally. The correct statement is: if we measure the position of
the quantum object, then we shall find it at one of the two positions with
some probability, but it would be incorrect to assume that it was located
at this position before we performed the measurement. Contrary to what
happens in classical physics, *a measurement does not reveal a pre-existing
reality*. When a police officer measures the velocity of a car on a highway
with a radar gun, this velocity existed before the measurement, which gives
the police officer the legitimacy to issue a speeding a ticket if the velocity
exceeds 130 km/h (in France, with fine weather). This is not the case in
quantum physics: an atom may well be in a state where its velocity is not
determined, for example in a state which is a linear superposition of two
states with different velocities. A measurement will give one of these two

velocities, but it does not mean that the atom possessed this velocity before the measurement.

> *The fact that a quantum object can be in a state of linear superposition (or simply superposition) of two or more quantum states is a fundamental characteristic property and, without any doubt, the most characteristic property of quantum physics.*

We do not have any intuitive experience of this *superposition principle* in everyday life, which is governed by the principles of classical physics. This is the reason why this principle is very surprising: it contradicts our usual view of the world, and in some sense it is "an inconvenient principle" — an expression borrowed from Al Gore's "inconvenient truth". This explains why debates are still going on today on the interpretation of this principle, a point which we shall return to in Chapter 10. For the time being, we must simply try to live with this principle in order to understand the physics in what follows.

1.4 Photons in the Mach–Zehnder interferometer

The results obtained with the Mach–Zehnder interferometer do not display anything extraordinary as long as we stay within the framework of classical wave theory and use a laser as a source of light. The situation becomes much more interesting if we use a single photon source instead of a laser, so that we must take into account the particle aspect of light. Single photons are emitted by a source at tagged instants of time, and knowing the length traveled in the interferometer, we can identify the coincidences between the emission of a given photon and its detection by D_X or D_Y. For simplicity, and in order to avoid any ambiguity, we shall assume that there is at most one photon traveling in the interferometer between emission and detection of a given photon. Let us begin first with a naive, but incorrect reasoning, because criticizing an incorrect reasoning is often instructive. A given photon has a 50% probability of choosing the red path at BS_1 and then a 50% probability of choosing after BS_2 a path parallel to the horizontal direction or the vertical direction, so that the final probability for this photon of hitting D_X is 25%, and that of hitting D_Y also 25%. Since the photon could have taken the blue path at BS_1, both detectors are triggered with a 50% probability. To summarize, if Prob_X and Prob_Y are the detection probabilities of D_X and D_Y respectively, we expect the following result

$$\text{Prob}_X = \frac{1}{2}, \qquad \text{Prob}_Y = \frac{1}{2}.$$

But this "obvious" result is very likely wrong, as can be understood from the following heuristic argument. Let us start from the classical optics description of the interferometer (Section 1.2) and let us reduce the laser intensity so that, on average, there is at most one photon traveling in the interferometer between emission and detection of a given photon. Then the detectors register isolated counts. The result from classical optics in Section 1.2, $I_X = I$ and $I_Y = 0$, leads us to predict $\text{Prob}_X = 1$ and $\text{Prob}_Y = 0$ if we want to recover classical optics in the limit of a large number of photons, but this result is at variance with that of our preceding probabilistic reasoning. Of course, we have explained that attenuated light is not a single photon source, and the experiment should be performed with an actual single photon source. In that case, experiment confirms the result derived from classical optics, which is typical of an interference phenomenon

$$\text{Prob}_X = 1, \qquad \text{Prob}_Y = 0.$$

As the correct result is that suggested by a wave picture, it means that we must reason on amplitudes and not on probabilities. Actually, the argument using wave amplitudes can be immediately transposed to one using probability amplitudes in order to derive the correct result.

The *first lesson* to be drawn from this discussion is that the photon displays in the same experiment particle properties, when it hits a detector in a well defined point in space (one never observes a fraction of a photon), and wave properties, since the fact that 100% of the photons trigger D_X is typical of constructive interference. A photon is neither a classical particle (a micro-billiard ball), nor a wave. It is an entirely new object, of which we have no good intuition, as our intuition is grounded on the observation of macroscopic phenomena which are described by classical physics. No classical description can do justice to the complexity of photon behavior: a photon does not behave like anything one has ever seen, it is a *quantum object*, so that some physicists have coined the word "quanton" to underline the unituitive aspect of quantum behavior.

The *second lesson* is that the role of light *intensities* is now played by *probabilities*: instead of the intensities I_X and I_Y, the relevant quantities are the *probabilities* Prob_X and Prob_Y. In a similar way, as we saw in § 1.3.3, the wave amplitudes a_X and a_Y become probability amplitudes whose square gives the probabilities, for example the probability of triggering detector D_X: $\text{Prob}_X = |a_X|^2$.

The *third lesson is that the notion of photon trajectory loses its meaning.* Indeed, assume that we can determine whether a given photon of our ensemble of single photons chose the blue path or the red one in Figure 1.3, or in other words, assume we have been able to tag the photon path. We shall see in Chapter 9 a possible way of doing so. Thanks to that tagging, we can divide the photons into two groups, the group which followed the blue path and that which followed the red one. Let us concentrate on the latter group: when a photon belonging to this group arrives on beamsplitter BS_2, it has a 50% probability of taking the X direction and a 50% probability of taking the Y direction, so that $\text{Prob}_X = \text{Prob}_Y = 1/2$, and there is no interference. A configuration in which interference is observed implies that one cannot answer the question: which is the path followed by the photon? This point is so important that it is worth giving a second version of the argument. In the presence of interference, we could be tempted to argue that one of the following two alternatives has been realized for a given photon.

(1) *Alternative 1.* The photon followed the blue path.
(2) *Alternative 2.* The photon followed the red path.

But if either of these alternatives had been realized, one would not observe interference, as we could divide the photons into two groups even in the absence of tagging. For both groups, no interference is possible, and thus *neither of the two alternatives has been realized.* In others words, the question "which is the path followed by a given photon?" does not have any meaning; we are simply not allowed to ask it. We can try an "explanation" like "the photon followed both paths at the same time", but this is only a way of eluding the problem. In fact the only correct statement is "the photon state is represented by a linear (or coherent) superposition of the two paths".

Box 1.4. The de Broglie–Bohm theory.

The above reasoning, popularized by Feynman and which can be found in a number of textbooks, contains a flaw. As we shall see in Section 10.4, in the de Broglie–Bohm theory and if we consider a neutron (or atom) interferometer, in order to limit ourselves to non-relativistic velocities, the neutron *does* follow one of the two paths. Assume that the blue path is blocked when the neutron follows the red path and vice-versa. It could happen, and it is the case in the de Broglie–Bohm theory, that blocking one path or not blocking it may influence the neutron propagation on the other path. However, such theories are rejected by a majority of physicists because they are non-local, in the sense that they imply instantaneous action at a distance (see Chapters 3 and 10).

1.5 The Mach–Zehnder interferometer revisited

We are going to complicate (slightly) the Mach–Zehnder interferometer by increasing the distance traveled by light along one of the two arms thanks to four additional mirrors. On the blue path, light travels an additional distance ℓ (Figure 1.6). If this distance is a multiple of the wavelength, $\ell = n\lambda$, n an integer, nothing is changed since we have the same wave just before BS_2, whether the additional mirrors are present or not: if the waves

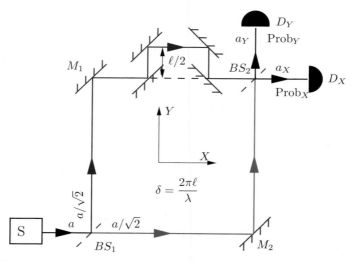

Figure 1.6.   Modified Mach–Zehnder interferometer. Light travels an additional distance ℓ on the blue path and the phase shift $\delta = 2\pi\ell/\lambda$.

issued from the red and blue paths were originally in phase, they remain so. If ℓ is of the form $\ell = (n + 1/2)\lambda$, the wave from the blue path will be out of phase with that of the red path before BS_2, and the intensity in D_X will vanish. In the general case, the intensity in D_X is given as a function of the angle $\delta = 2\pi\ell/\lambda$ by (Appendix A1.4)

$$I_X = \frac{I}{2}(1 + \cos \delta), \quad I_Y = I - I_X = \frac{I}{2}(1 - \cos \delta). \quad (1.3)$$

δ is the phase shift introduced in § 1.1.1, and it is a measure of the delay suffered by light owing to the additional mirrors. For the benefit of the reader whose high school math is a bit rusty, I have plotted the function $\cos \delta$ in Figure 1.7. One obtains $I_X = I, I_Y = 0$ if δ is a multiple of 2π, $\delta = 2\pi n$, n an integer, while the opposite case $I_X = 0, I_Y = I$ is obtained when $\delta = (2n + 1)\pi$. In general, the intensity is split in the two directions and the repartition varies as a function of δ according to the law (1.3). This dependence with respect to δ is the signal for interference.

The case of single photons is easily deduced from that of light waves, as the rules for combining probability amplitudes are the same as those for combining wave amplitudes. As in the physics of waves, these amplitudes can interfere constructively, destructively, or in an intermediate way. The result of an experiment performed by Aspect *et al.* [1989] with a single photon source is given in Figure 1.8. Each detector registers isolated counts, a given photon triggering either D_X or D_Y. When the number of photons increases, we observe the progressive building of an interference pattern

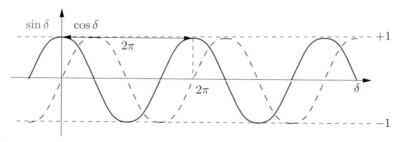

Figure 1.7. The function $\cos \delta$ (full line, green). This functions repeats itself with a period 2π: the interval between two consecutive maxima is 2π. The cosine function lies between -1 and $+1$: $-1 \leq \cos \delta \leq +1$. Its value is $+1$ for $\delta = 0, 2\pi, 4\pi, \ldots$ and -1 for $\delta = \pi$, $3\pi, 5\pi, \ldots$. The sine function $\sin \delta$ is drawn with a red dashed line.

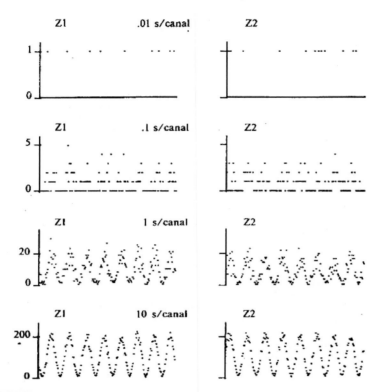

Figure 1.8. The progressive construction of an interference pattern built photon after photon by detectors $D_X(Z_1)$ and $D_Y(Z_2)$. The phase shift δ is given in abscissa. One remarks that the two detectors give complementary signals in $(1 + \cos\delta)/2$ and $(1 - \cos\delta)/2$. The time spent for registering photons for a given value of δ varies between 0.01 s (upper figure) and 10 s (lower figure). Courtesy of Alain Aspect and Philippe Grangier.

which reproduces the cosine dependence of equation (1.3). Indeed, in order to reproduce the results of classical optics when the number of registered photons becomes large, the photons which follow the horizontal path after BS_2 must trigger D_X with a probability $\text{Prob}_X = (1 + \cos\delta)/2$, and those which follow the vertical path must trigger D_Y with a probability $\text{Prob}_Y = (1 - \cos\delta)/2$ respectively. One checks that the total detection probability is one, $\text{Prob}_X + \text{Prob}_Y = 1$, as it should be: since the photons are not absorbed when they travel in the interferometer, a photon which enters the interferometer must trigger either D_X or D_Y.

1.6 Quantum particles

1.6.1 de Broglie waves

There is little doubt that photons behave in a strange way, but they are not alone to do so! All particles of microscopic physics: electrons, protons, neutrons, atoms, molecules ... share this surprising behavior. But why should we not also consider as quantum objects proteins, viruses, billiard balls or even elephants? We do not have a completely definite answer to the question: is there a limit of size for quantum behavior? However, we have some hints of a possible answer which will be discussed in Chapter 9.

Let us now give further details of the quantum description. If we say that a quantum particle shares some common properties with a photon, it implies that this particle possesses both particle and wave properties. For an atom, the particle aspect is obvious: we can imagine an atom as a micro-billiard ball whose diameter is on the order of one tenth of a nanometrer (1 nm $= 10^{-9}$ m, a billionth of a meter). But what about wave properties? In 1923, Louis de Broglie stated his famous hypothesis that the wavelength of a massive quantum particle is related to its mass m and to its velocity v by the relation

$$\lambda = \frac{h}{mv}. \tag{1.4}$$

λ is *de Broglie's wavelength*. In fact, this equation is only valid for velocities which are small with respect to the velocity of light, $v \ll c$. For such low velocities, the *momentum p* is given by $p = mv$, and one can rewrite equation (1.4) as $\lambda = h/p$. This latter form is then generally valid, provided one uses for the momentum the formula (7.2) given by special relativity. In the photon case, momentum is related to energy E by $p = E/c$, and combining the Planck–Einstein relation (1.1) and the de Broglie relation (1.5), we recover $\lambda = h/p$.

The de Broglie hypothesis implies that we should observe for any quantum particle phenomena characteristic of waves, that is, diffraction and interference. In fact, diffraction and interference have been observed for

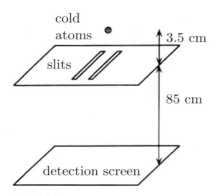

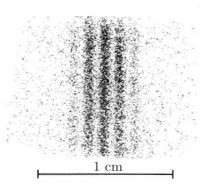

Figure 1.9. Interference with cold atoms. The cold atoms fall under the influence of gravity on the two slits and they spread out on the screen building an interference pattern. One should observe that the interference pattern is formed of *isolated* impacts of atoms on the screen. From J-L Basdevant and J. Dalibard, *Mécanique quantique*, Éditions de l'École Polytechnique, Palaiseau, experiment by Shimizu *et al.* [1992].

a large variety of particles: electrons, neutrons, atoms, molecules. Let us limit ourselves to three examples.

1. It is common practice in laboratories to work with Mach–Zehnder interferometers using neutrons or atoms, which are dedicated to precision experiments.

2. A spectacular Young's slit experiment has been performed with cold atoms (Chapter 5). The atoms fall under the influence of gravity on the two slits and form an interference pattern on a screen (Figure 1.9).

3. The gyrolasers used for inertial navigation in planes, and which have replaced gyroscopes, are based on the same principle as the Mach–Zehnder interferometer. A rotation with an angular velocity ω along an axis perpendicular to the plane of the interferometer induces a phase shift $\delta = 2\pi\omega\nu S/c^2$, where S is the area bounded by the arms of the interferometer and ν is the light frequency. The interference pattern depends on δ, and this allows one to deduce the angular velocity ω. Instead of photons, one can use in the interferometer neutrons or atoms. The sensitivity is then considerably increased, but it is not easy to work with neutrons or atoms in planes!

Box 1.5. The Sagnac effect.

The effect which is used in the gyrolaser is the Sagnac effect. When the interferometer rotates along an axis perpendicular to its plane, the beam which follows the sense of rotation arrives on BS_2 after that which takes the other path. For an angular velocity which is small enough, the difference between the two times of arrival is $\Delta t = 2\omega S/c^2$, hence the phase shift δ. If one uses in the interferometer neutrons or atoms, then $\delta = 4\pi m\omega S/h$. It is amusing to notice that the simplified formula for photons given just above can be derived by using the ether theory, discredited by Einstein. Actually, Sagnac had seen in this effect a confirmation of the ether theory!

1.6.2 Rules of quantum physics

Quantum objects behave in a strange way, with their dual wave and particle properties, but there is nevertheless a simplification. As Feynman puts it "Quantum objects are completely crazy, but at least they are all crazy in the same way", and they all obey the same rules. At this point, it may be useful to make a first inventory of these rules, sometimes called "postulates of quantum physics", or "axioms of quantum physics", but I prefer to use the more neutral term "rules". It is clearly impossible to give a "proof" of these rules, and, as far as we know, these rules do not derive from concepts which would be still more fundamental. As of today, it is only possible to state these rules *a priori*, as far as they are not in contradiction with experiment. They are grounded on the notion of *process*, defined by an initial state and a final state. For example, in the case of the Mach–Zehnder interferometer, the initial state corresponds to the photon entering the interferometer at BS_1 and the final state to its detection at D_X.

Rule number 1. To every process defined by an initial state and a final state, there corresponds a probability amplitude a. Example: the number a_X in Figure 1.3 is the probability amplitude that a photon entering the interferometer (initial state) be detected by D_X (final state).

Rule number 2. The probability of observing a process is the square a^2 of its probability amplitude, Prob $= a^2$, or more exactly, if we remember that a is in general a complex number, its modulus squared, Prob $= |a|^2$. This is the so-called *Born's rule*. Example: the probability that D_X is triggered is $\text{Prob}_X = |a_X|^2$. It often happens that we do not bother to normalize the probabilities in such a way that the sum over all possible cases is one, so that the probability is only proportional, and not exactly equal, to the amplitude squared.

Rule number 3 (superposition principle). If there exist two processes starting from the same initial state and leading to the same final state, with respective probability amplitudes a_1 and a_2, then the total amplitude a is obtained by adding the two amplitudes: $a = a_1 + a_2$. It follows that the probability depends not only on the squares a_1^2 and a_2^2, but also on the phase shift δ between the two amplitudes, and the total probability is given by (Appendix A.1.3)

$$\text{Prob} = a_1^2 + a_2^2 + 2|a_1 a_2| \cos \delta. \qquad (1.5)$$

If we use complex numbers, this relation is equivalent to $|a|^2 = |a_1 + a_2|^2$. It is clearly this property which is at the origin of interference. Example: in the Mach–Zehnder interferometer, we must add up the two amplitudes coming from the blue path and the red path. The rule is applicable only if it is impossible to distinguish between the two paths. On the contrary, if the experiment allows us to tell between the two paths, then we must add *probabilities*: $\text{Prob} = \text{Prob}_1 + \text{Prob}_2 = a_1^2 + a_2^2$, and interference is destroyed. It is sufficient that the distinction be possible *in principle*, but it is in no way necessary that it is effectively observed. It could even happen that today's technology be completely unable to make the distinction, even though it is theoretically conceivable: this does not influence the rules. In order to destroy interference, it is enough that there remains some trace of the path in the environment, even if we do not know how to make use of it. To summarize: it is the experimental configuration which dictates the correct application of the superposition principle. This statement will be checked in the next section with the example of the delayed choice experiment.

1.7 Delayed choice and interaction-free measurement

We shall conclude this chapter with two further illustrations of what can be done with a Mach–Zehnder interferometer: the delayed choice experiment and interaction-free measurements. This section is not essential and may be skipped in a first reading. Let us begin with the first case and describe a concrete realization of an experiment first proposed by Wheeler, assuming that the second beamsplitter BS_2 in Figure 1.3 can be moved at will: it can be present as is sketched in Figure 1.3, or it can be moved out of the interferometer. If the beamsplitter is present, then we observe interference, $\text{Prob}_X = 1$, $\text{Prob}_Y = 0$, and if it absent, $\text{Prob}_X = \text{Prob}_Y = 1/2$. Furthermore, let us assume that the beamsplitter can be moved very fast, so that it may be introduced or removed after the photon has already crossed the first beamsplitter BS_1, for example when the photon is in the neighborhood of one of the mirrors. This experiment is called delayed choice because the decision of introducing the second beamsplitter or not is taken *after* the photon has crossed the first one and travels in the interferometer. One could naïvely think that there is no interference, whether the second beamsplitter is present or not, because after having crossed the first beamsplitter, the photon must have chosen either the blue path or the red path. However, after the second beamsplitter has been introduced, the two paths cannot be distinguished, and rule number 3 implies that one should observe interference. The experiment was performed in 2007 by a team at the Institut d'Optique/ENS Cachan (Figures 1.10 and 1.11), using an interferometer slightly different from that of Figure 1.3. In this interferometer, the distance

Figure 1.10. Photography of part of the apparatus, the single photon source, used in the delayed choice experiment performed by Jacques *et al.* [2007]. Courtesy of Jean-François Roch.

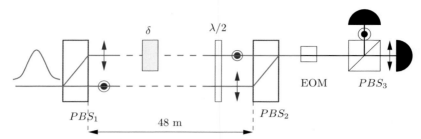

Figure 1.11.   Sketch of the delayed choice experiment. A single photon enters the interferometer at a first polarizing beamsplitter (PBS, see the next chapter for details). It "follows" one of the two paths according to its polarization state, vertical on the upper path and horizontal in the lower path. A half wave plate swaps the polarizations, so that the two paths can be recombined by a second PBS, PBS$_2$. A third PBS, PBS$_3$, after the interferometer, allows one to distinguish between the two polarization states and thus to tag the path followed by the photon: there is no interference, the polarization plays the role of a marker of the path. An electro-optic modulator EOM, if activated, rotates the polarization by $45°$, which erases the information on the path. If the EOM is activated, one recovers the interference situation.

traveled by the photon is 48 m, and the corresponding time is $0.16\,\mu$s. No mechanical device is able to move a beamsplitter in such a short interval of time, and the experimentalists had to use an electro-optic modulator, whose action is equivalent to moving a beamsplitter or not. The device is sketched on Figure 1.11, and the experimental results vindicate the theory beyond any doubt: interference is observed if and only the beamsplitter is present, and clearly the notion of a photon trajectory loses any meaning in the presence of interference.

This experiment seems to vindicate Bohr's point of view, according to which quantum objects have no existence independent of the experimental apparatus used to observe them (Chapter 10). When the second beamsplitter is present, the photon which travels in the interferometer is simply *not* the same object as when the beamsplitter is absent. This point of view is grounded on the distinction between a classical (macroscopic) world and a quantum (microscopic) world. In other words, it relies on the existence of a frontier between a classical world, the only world to which we have direct access, and a quantum world, which we can only know through macroscopic devices. However, as we shall see in Chapter 9, there are good reasons to question the existence of such a classical-to-quantum frontier.

Let us now turn to interaction-free measurement, assuming that an obstacle is put on that part of the red path oriented along the Y axis

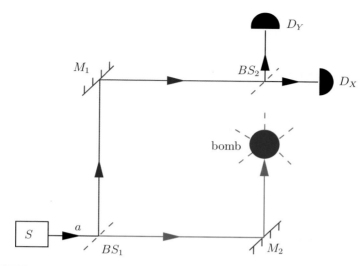

Figure 1.12. Device for an interaction-free measurement. S = single photon source.

(Figure 1.12), which prevents the photon from reaching the second beam-splitter. The interferometer has been set exactly as in Section 1.2, so that $\text{Prob}_X = 1$, $\text{Prob}_Y = 0$ in the absence of an obstacle: all the photons are detected by D_X. In order to make the argument more dramatic, let us assume that the obstacle is a bomb which explodes as soon as it is hit by a single photon. It is thus impossible to see the bomb without catastrophic consequences. We do not know in advance whether the bomb is present or not, and there are three possibilities: if the bomb is present, the photon may follow either the red path or the blue path, and if it is absent, there is interference.

1. The bomb explodes. Then one concludes that the bomb was present, but neither of the detectors is triggered since the apparatus was destroyed!
2. Detector D_X is triggered: in that case, either the bomb was present and the photon followed the blue path, or it was absent, and we had an interference configuration.
3. Detector D_Y is triggered.

This last case is the interesting one, since one can conclude that the bomb was present and the photon followed the blue path. Then we have been able to detect the presence of the bomb without interacting with it: this is an example of an interaction-free measurement. We have been able to probe a region of space without interacting directly with it.

1.8 Further reading

A good introduction at an elementary level is Scarani [2006]. See also Hey and Walters [2003] Chapters 1 to 4, the first chapters of Feynman *et al.* [1965], vol. III, and, at a slightly more advanced level, Scarani *et al.* [2010]. There exist dozens of textbooks on quantum mechanics, among which I recommend: Cohen-Tannoudji *et al.* [1977], Peres [1993] and Ballentine [1998]. The experiment described in Figure 1.5 was performed by Jacques *et al.* [2005] and the delayed choice experiment by Jacques *et al.* [2007].

2

Secure Communications

The first practical application of quantum physics examined in this book is quantum cryptography. Quantum cryptography is a relatively recent invention (it dates back from the mid 1980s) but I chose it because it allows me to illustrate the fundamental principles with a minimum number of intermediate steps. I shall begin with a short summary of classical cryptography, reviewing briefly the two systems which are currently used today: the secret key system and the public key system. Quantum cryptography is not a new method for dissimulating the meaning of a message, but it allows one to be certain that no spy has accessed it. There exist many quantum cryptography protocols and various experimental devices have been proposed for implementing them. The simplest device is based on polarization, a concept which will be introduced first in the case of of light polarization, and then in that of photon polarization. The use of photon polarization gives the simplest implementation of the protocol proposed in 1984 by Bennett and Brassard, which is known by the acronym formed with their initials, the BB84 protocol.

2.1 Classical cryptography: Secret key and public key

Cryptography, the art of transmitting to an authorized party a message whose meaning cannot be understood by an adverse party, is almost as old as writing. Such a message is called encrypted, or sometimes encoded (*cf.* "The Da Vinci Code"). The first modern encryption is probably due to Julius Cæsar, around 50 BC, and is called the substitution method: to each letter of the alphabet is substituted another letter, for example, the following letter in the alphabet. The message "THE LEGIONARY VOREMUS..." (see the BBC series "Rome") becomes "UIF MFHJPOBSZ WPSFNVT...". It is well-known that such a code is very easy to crack, even if one uses a more sophisticated method than a one letter shift: a code breaker exploits the frequency of appearance of letters or groups of letters, in particular double letters, in a given language. This encryption was considerably improved around 1550 by Blaise de Vigénère, who had the idea of using a key which allows one to encode the same letter by a different one, according to its position in the message: for example, the same letter, A, could be encoded by E, X or H. This encryption was able to resist for three centuries the efforts of code breakers, until Charles Babbage finally cracked it around 1850.

The next improvement was automatic encoding. In 1918, Arthur Schoebius built an electromechanical machine — electronics was in its infancy at that time! — which he christened ENIGMA. With this machine, the text of the original message, or plaintext, is typed directly on a keyboard, and light bulbs are switched on to give the encoding of each letter, which is obtained by a system of rotors whose arrangement is determined by an encryption key. The receiver deciphers in analogous manner: the encrypted text is typed on a keyboard and light bulbs give for each letter the corresponding letter in the original message. This machine was improved by the Nazi army and widely used by it during World War II, especially to communicate with its submarines. In order to protect allied convoys from the German submarines, a dramatic race was led with success by the code breakers of Bletchley Park, under the direction of the mathematical genius Alan Turing. The number of possible keys is about 10^{16}, so that even with the computing power available today, it would be impossible to crack ENIGMA by brute force.

Until the 1970s, cryptography was almost exclusively reserved to the army and to diplomacy. With the explosion of electronics and optical fiber communication, cryptography has become part of our everyday life: cryptography is mandatory for communications with banks, commercial transactions, purchases on the Internet, tax returns and so on. In the acronym https://, the "s" indicates that the site is secured, and — at least in principle — your visa card number or your tax record is not directly visible on the net.

As of today, there exist two main encryption methods. The first one is called secret key encryption, or symmetrical encryption, and the second one public key encryption, or asymmetrical encryption. The classic example of *secret key encryption* was first described by Vernam in 1926. The sender and the receiver must share a key, which is unknown to anyone else, hence the name secret key. The key is a random sequence of 0s and 1s, and the message is first translated into binary language, as a series of 0s and 1s. The crypted message is obtained by adding the key to the message, using the so-called modulo 2 addition without carry-over, the "exclusive or" of computer scientists, which is denoted $\oplus$ and is defined by the following rules

$$0 \oplus 0 = 0, \quad 0 \oplus 1 = 1 \oplus 0 = 1, \quad 1 \oplus 1 = 0.$$

This encryption is illustrated below with a simple example

$$\begin{aligned}
\text{message} \quad &: \quad 1\,0\,0\,1\,1\,0\,1\,0\,0 \\
\text{key} \quad &: \quad 1\,1\,0\,1\,1\,1\,1\,0\,1 \\
\text{encrypted message} \quad &: \quad 0\,1\,0\,0\,0\,1\,0\,0\,1
\end{aligned}$$

Deciphering is done by adding simply the crypted message and the key

$$\begin{aligned}
\text{encrypted message} \quad &: \quad 0\,1\,0\,0\,0\,1\,0\,0\,1 \\
\text{key} \quad &: \quad 1\,1\,0\,1\,1\,1\,1\,0\,1 \\
\text{original message} \quad &: \quad 1\,0\,0\,1\,1\,0\,1\,0\,0
\end{aligned}$$

Claude Shannon proved that the encryption is perfectly secure, provided:

1. The key is truly random.
2. The key is used only once: this is the famous one-time pad condition.
3. The key is as long as the message.

The weakest link of secret key encryption lies in point (2) above: the key must frequently be shared between partners, and this is a risky process. For example, the partners meet at the same location to exchange keys, which may be very inconvenient, or trust a third party to carry the key from one partner to the other, or store the keys in a safe, and so forth. In all cases the process is either complicated or risky (or both!). For this reason, one often prefers to use an encryption founded on a completely different principle, the *public key encryption*, where the key can be issued publicly, for example on the Internet, or transmitted by phone. The standard public key encryption, called *RSA encryption* (an acronym formed with the initials of its inventors, Ronald Rivest, Adi Shamir and Leonard Adleman in 1977) is based on the difficulty of factoring an integer N into primes, that is, to write it as a product of primes, while the inverse operation is easy: even without the help of a pocket calculator, one will obtain in a few seconds the result $137 \times 53 = 7261$, but given 7261, it will take some time to obtain the prime factors 137 and 53. With the fastest algorithms which are known today, the time necessary to factor an integer N increases as $\simeq \exp[1.9(\ln N)^{1/3}(\ln \ln N)^{2/3}]$, where $\ln N$ is the logarithm of N, roughly the number of digits which are needed to write N. At present, it is possible to factor into primes a number of about 250 digits in a reasonable amount of time, and if one month is necessary for this factoring, then the same computer will need some 100 000 years to factor a number of 400 figures.

In the public key encryption system, the receiver, conventionally called Bob, sends publicly to his partner, Alice, a very large number $N = pq$ which is the product of two primes p and q, as well as another integer c: it is the receiver who sends the key! The two integers N and c are sufficient for Alice to encrypt the message, but one must know the two primes p and q in order to decipher it (see Appendix A.2.1). However, there is no mathematical proof that it will not be possible to discover faster algorithms to factor a large integer into primes, and if quantum computers see the light of the day, then factoring will become a "tractable" problem: to go from 200 to 400 digits, we need only multiply the computing time by a factor of 8.

Box 2.1. Algorithmic complexity.

The theory of *algorithmic complexity* allows us to classify problems into two main categories. On the one hand, there are "tractable problems", those whose solutions on a computer is such that the number of operations grows only moderately with the size of the problem. In the case of factoring, the size of the problem is simply the number of digits of the integer which one plans to factor. On the other hand, there are "intractable" problems, those whose solution requires a number of operations which grows very fast with the size of the problem, so that the solution cannot be found in practice, although there exists in principle an algorithm to solve the problem. It is not known whether factoring into primes is a "tractable" or an "intractable" problem if one uses a classical computer. With a quantum computer, factoring into primes becomes "tractable": if the size of the problem is doubled, the necessary time increases by a factor of 8 only. This point will be explained in more detail in § 8.3.

2.2 Light polarization

Classical cryptography relies either on the use of a secret key whose secure transmission may be problematic, or on mathematical properties which are not proven. RSA encryption would immediately collapse if factoring into primes turned out to be a "tractable" problem, hence the idea of using quantum cryptography which, at least in principle, is completely secure. Quantum cryptography is a popular expression, but it is somewhat misleading. Indeed, one does not attempt to encrypt a message using quantum physics, but the idea is to establish a key which can be used later for a classical protocol, while being certain that this key remains secret: the correct expression is then "Quantum Key Distribution" (QKD). Thus, quantum cryptography is meant to solve the problem of secure sharing of a key to be used later on in a classical protocol, and the safety of the procedure relies uniquely on the laws of quantum physics. In order to explain the basic principles, we shall describe the protocol proposed in 1984 by the computer scientists Charles Bennett (IBM) and Gilles Brassard (Montreal), which is known as the BB84 protocol. Furthermore we shall limit ourselves to the simplest physical implementation of the protocol, that based on polarization. *Light polarization* is a familiar phenomenon, whose origin lies in that light oscillates in a plane perpendicular to its direction of propagation. Let us consider a light beam propagating along a horizontal direction. A *polaroid* is a material with a preferred axis: if this axis is oriented vertically, it transmits only light which oscillates along this axis and the light which leaves the polaroid is vertically polarized (Figure 2.1a). If we place after the first polaroid a second one whose axis may lie along different directions perpendicular to the direction of propagation, we observe the following results.

1. If the axis of the second polaroid is oriented vertically, then 100% of the light leaving the first polaroid is transmitted.
2. If the axis is horizontal, then the second polaroid does not transmit any light at all: two polaroids whose axes are at right angle stop any light beam. Polaroid sunglasses work according to this principle: when they are normally oriented on the nose of their owner, they transmit uniquely vertically polarized light and stop horizontally polarized light. This allows one to eliminate bothersome reflections, for example, in the

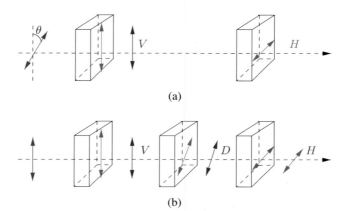

(a)

(b)

Figure 2.1. A light ray going through a polaroid. (a) The polarization of the incident light makes an unknown angle θ with the vertical axis. Part of the light intensity is stopped by the vertically oriented polaroid. The part which leaves the polaroid is vertically polarized (blue), and it is stopped by the horizontally oriented polaroid. (b) The first polaroid transmits 100% of the incident light which is vertically polarized. The second polaroid is oriented along an axis which makes a 45° angle with the vertical direction, and the light which leaves the intermediate polaroid (blue) is polarized along this direction; 25% of the incident light (purple) is transmitted by the last polaroid, and it is horizontally polarized.

case of light reflected by snow, because the reflected light is mainly horizontally polarized.

3. If the axis of the second polaroid makes a 45° angle with the vertical direction, 50% of the light leaving the first polaroid is transmitted. This implies that if one inserts between two polaroids at right angle a polaroid whose axis makes a 45° angle with the vertical direction, then 25% of the light is transmitted, and this light is horizontally polarized (Figure 2.1b).

Instead of a polaroid which transmits light oscillating along one direction and stops light oscillating in the perpendicular one, it is often more convenient to use a *polarizing beam splitter*, or PBS, which divides an incident beam into two secondary beams according to their polarization (Figure 2.2). If the PBS axis is vertically oriented, then one of the secondary beams is vertically polarized: $\updownarrow$, or V. The other beam is horizontally polarized: $\leftrightarrow$, or H. If one filters incident light with a polaroid whose axis makes a 45° angle with the vertical direction, a PBS with a vertical axis will send 50% of the light into the vertically polarized beam, and 50% into the horizontally polarized one. Note that if the direction of the (linear) polarization of a

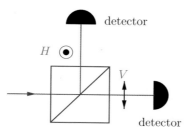

Figure 2.2.   A polarizing beam splitter divides an incident light beam into two sub-beams with orthogonal polarizations. The polarization of the sub-beam which follows the horizontal direction lies in the plane of figure (vertical or V polarization), while the polarization of the vertical sub-beam is perpendicular to this plane (horizontal, or H polarization).

light beam is unknown, it is possible to determine it by measuring the ratio of the transmitted intensity to the incident one using a polaroid (or a PBS) with a variable orientation: when the ratio is one, the polaroid axis gives the direction of the incident polarization.

2.3 Photon polarization

2.3.1   The no-cloning theorem

We do not encounter any unexpected problem of principle when describing light polarization, which is nothing other than a familiar wave phenomenon: vibrations in solids, or sound waves, may also be polarized. This is not the case for photon polarization, for which we shall encounter the weirdness of quantum physics. As we saw in the first chapter, we have known since Einstein (1905) that light exhibits particle properties: there exist "light particles" called photons. We know today how to build reasonably efficient single photon sources, where a photon can be tagged by the instant when it is produced. Exactly the same as light, single photons may be polarized, vertically, horizontally, or at 45°, or along any direction perpendicular to their direction of propagation. In addition to the vertical (V) and horizontal (H) polarizations, we shall need in what follows polarizations oriented at 45° (D, for diagonal) and −45° (A, for anti-diagonal).

Let us, for example, examine photons with a diagonal polarization D which may be obtained when single photons are transmitted by a polaroid oriented at 45° with respect to the vertical axis, the intermediary polaroid of

Figure 2.1b. If we analyze the polarization of these photons by letting them pass through a polaroid with vertical axis, they are transmitted with a 50% probability, and the same results hold for a polaroid with horizontal axis, in agreement with the sketch of Figure 2.1b drawn for polarized light. But that does *not* mean that the photons incident on the polaroid are polarized along the vertical direction with a 50% probability, or along the horizontal direction with a 50% probability. Exactly the same as the photon in Figure 1.4 was in a linear superposition of states localized in the neighborhood of two different positions, a photon polarized along the D axis is in a *linear superposition* of states polarized along the V and H axes, which we shall call the V and H states of polarization, while D will be the superposition $D = (V + H)/\sqrt{2}$. With this notation, $1/\sqrt{2}$ is a *probability amplitude*, whose square $1/2$ gives the 50% *probability* that a photon in a state of polarization D goes through a polaroid oriented vertically or horizontally, in agreement with the preceding discussion. The state of polarization A is represented by $A = (V - H)/\sqrt{2}$, which emphasizes that the *signs* of probability amplitudes strongly matter, and not only their squares.

We may now ask the following question: is it possible to determine the polarization of an incident photon when the *only* available information is that this photon is polarized either vertically (V) or horizontally (H), or, on the contrary, either along the $+45°$ direction (D) or the $-45°$ direction (A)? Let the photon go through a polaroid with a vertically aligned axis. If this photon is transmitted, this means that it was certainly not polarized horizontally, but does not exclude the possibility that it was polarized at $\pm 45°$, in which case it would have had a 50% probability of being transmitted by the vertically oriented polaroid. But, in order to check this latter hypothesis, we would need a second identical photon to analyze its polarization with a polaroid oriented at $+45°$: this would allow us to determine whether the photon was a D-photon or an A-photon. This is not possible because we had only one photon at our disposal. This argument shows that, in order to determine the polarization of a photon, we need to be able to duplicate it in order to perform an analysis, either along the vertical/horizontal directions, or along the $\pm 45°$ directions.

In classical physics, there are no limitations to the possibility of copying information: we make a daily use of this property when we photocopy a document or when we download a document on the Internet and copy it to the hard drive of our computer. This is not the case in quantum physics, where the copying possibilities are constrained by a fundamental theorem, which is a direct consequence of the superposition principle: the *quantum no-cloning theorem*. In order to prove this theorem in a simple case, but the result is quite general, we shall limit ourselves to four possible orientations for the polarization, corresponding to two different *configurations* (or orthonormal bases in a mathematical language, see Box 2.2): vertical/horizontal, or $\{VH\}$ and $\pm 45°$, or $\{DA\}$. We are going to show that it is possible to copy, or clone, the polarization state carried by photons if they are *all* in the configuration $\{VH\}$, *or all* in the configuration $\{DA\}$, but a device able to clone photons in the $\{VH\}$ configuration will not be able to do so for photons in the configuration $\{DA\}$, and vice-versa. Let us assume that there exists a quantum cloning machine, a Q-photocopier, able to duplicate the polarization states V and H carried by a photon. We thus have a Q-photocopier with a blank page, a polarization state X, on which we wish to "print" the state to be duplicated. This Q-photocopier has the following action for initial states V and H

$$VX \Rightarrow VV, \quad HX \Rightarrow HH.$$

From these rules, we deduce the action on the state D

$$DX = \frac{1}{\sqrt{2}}(V + H)X \Rightarrow \frac{1}{\sqrt{2}}(VV + HH).$$

But this is not the state we want, as duplicating D would give

$$DX \Rightarrow DD = \frac{1}{\sqrt{2}}(V + H)\frac{1}{\sqrt{2}}(V + H) = \frac{1}{2}(VV + VH + HV + HH).$$

This puts and end to the proof: the Q-photocopier cannot duplicate correctly the D state.

To summarize: if we know in advance the configuration of the incident photons, $\{VH\}$ or $\{DA\}$, a PBS with a suitable orientation will be able to determine the state of polarization: in the $\{VH\}$ configuration, the PBS must be oriented as in Figure 2.2, and in the $\{DA\}$ case, we must rotate the PBS by a $45°$ angle around the direction of propagation. If we know that the photons belong to one and only one configuration, but if this configuration is unknown, then we can determine their polarization by cloning a large enough number of photons and by analyzing their polarization. If we do not have any prior information of this kind, it is impossible to determine the polarization. We may then draw an interesting consequence from this analysis: it is impossible to determine an unknown quantum state.

Box 2.2. Orthonormal bases.

The mathematical term for "configuration" is *orthonormal basis*. As is explained in Appendix A.2.2, we may associate to the polarization states V and H two vectors noted $|V\rangle$ and $|H\rangle$ which form an orthonormal basis of a complex two-dimensional space

$$\langle V|V\rangle = \langle H|H\rangle = 1, \quad \langle V|H\rangle = 0,$$

where $\langle X|Y\rangle$ denotes the scalar product of two vectors $|X\rangle$ and $|Y\rangle$. The orthonormal basis $\{|V\rangle, |H\rangle\}$ corresponds mathematically to the configuration $\{VH\}$. The vectors $|D\rangle$ and $|A\rangle$

$$|D\rangle = \frac{1}{\sqrt{2}}(|V\rangle + |H\rangle), \quad |A\rangle = \frac{1}{\sqrt{2}}(|V\rangle - |H\rangle)$$

correspond mathematically to the $\pm 45°$ polarizations. These vectors are linear superpositions of the vectors $|V\rangle$ and $|H\rangle$, and they also form an orthonormal basis.

2.3.2 Encoding messages using photon polarization

The transmission of a message, be it encrypted or not, can be implemented by using the two orthogonal polarization states of a photon, for example, V ($\updownarrow$) and H ($\leftrightarrow$). A bit of information can take values 0 or 1, and since the photon polarization in a definite configuration, say $\{VH\}$, can take two values, we may by convention assign the value 0 to the vertical polarization and the value 1 to the horizontal one: each photon carries a bit of information. Once the message, encrypted or not, is written in binary language as a sequence of 0s and 1s, it appears, for example, as the sequence 1001110. It is encoded by Alice thanks to the polarization sequence *HVVHHHV* which she sends to Bob generally through an optical fiber. Using a PBS as in Figure 2.2, Bob is able to distinguish between the vertically polarized and the horizontally polarized photons: the vertically polarized photons trigger one of the detectors, and the horizontally polarized photons trigger the other one, so that he can read the message. Let us observe that if a spy, conventionally called Eve (for eavesdropper), taps the line, measures the

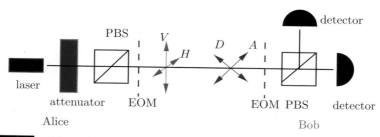

Figure 2.3. Schematic depiction of the BB84 protocol implemented with polarized photons. PBS = polarizing beamsplitter. EOM = electro-optic modulator. The photons are polarized either in the configuration {VH} (red arrows), or in the configuration {DA} (green arrows).

photon polarizations thanks to a PBS and sends Bob photons with polarizations identical to the initial ones, no strategy will be able to tell Bob that the line was tapped. The same statement is true for any transmission device which operates classically: if Eve is careful enough, she cannot be detected, because classical copying leaves no trace. Finally, if Eve does not know beforehand the polarization configuration used by Alice, but if she knows that this configuration is unique, she can measure the polarization by cloning the photons.

2.4 The BB84 protocol of quantum cryptography

2.4.1 The quantum protocol

It is at this point that quantum physics rescues Alice and Bob, by offering a strategy which allows them to be sure that the line was not tapped. We first describe the quantum part of the protocol. The goal is to establish a key which will be used later on for a classical encryption method where a secret key is required, and where the key needs to be replaced frequently. In order to exploit the no-cloning theorem, Alice randomly chooses her photons in two configurations, {VH} and {DA}, so that the photons cannot be cloned. Thus she sends Bob four types of photon: vertically polarized V ($\updownarrow$) and horizontally polarized H ($\leftrightarrow$), as in the preceding section, and polarized along axes oriented at $\pm45°$: A ($\nwarrow$) and D ($\nearrow$), corresponding respectively to the values 0 and 1 of the bits. In a similar way, Bob analyzes

Alice's polarizations	↕	↔	↗	↕	↗
sequence of bits	0	1	1	0	1·
Bob's configurations	+̷	✕	+̷	+̷	✕
Bob's results	0	0	1	0	1
bits for the key	0			0	1

Figure 2.4. Example of photon exchange between Alice and Bob: the green columns are kept to establish the key, the red columns are rejected.

the photons sent by Alice thanks to analyzers using the same configurations: vertical/horizontal {VH} and ±45° {DA}. Once more, the choice of configurations is made randomly, without taking into account any information about Alice's photons, whose polarizations at this stage are *unknown to Bob*. To perform the polarization analysis, Bob could use a PBS oriented in a random way either vertically or at 45° from the vertical axis. However, in practice, it is not a good idea to make fast rotations of the PBS, and one rather uses an electro-optic modulator, in which an electric signal can be tuned to rotate the polarization by ±45°. Then the ensemble PBS+detectors is fixed. Figure 2.4 gives an example of photon exchanges between Alice and Bob: Bob registers 0 if the photon polarization is ↕ (V) or ↖ (A), and 1 if the polarization is ↔ (H) or ↗ (D). Having registered a large enough number of photons, Bob announces publicly the sequence of configurations he has used, but not his results. Alice compares her own sequence of configurations to Bob's and gives him, again publicly, the list of configurations which coincide with his (green columns). The bits which correspond to incompatible configurations (red columns) are thrown away, and for the remaining bits, Alice and Bob can be certain that the values of these bits are identical: these are the bits which are used to establish the key, as Alice and Bob can be confident that they own an identical random series of 0s and 1s, which is unknown to a possible spy: Alice and Bob may have leaked their configurations to an adverse party, but not their results! One observes that the process eliminates roughly half of the bits sent by

Alice. One could think of keeping all the bits by storing the photons before making public the sequence of configurations, and reading the polarizations only afterwards. Unfortunately, the storage of single photons will remain for a long time a theoretician's fantasy.

2.4.2 The classical part of the protocol

The exchange of polarized single photons represents the quantum part of the protocol. But Alice and Bob still have to make sure that the line was not tapped and that their series of bits can be used without risk. In order to do it, they choose at random a common sub-ensemble of their key and compare it publicly. The consequence of an interception by Eve would be a reduction of the correlations between the values of their bits. Suppose, for example, that Alice sends a vertically polarized photon. In the simplest, but not the most efficient attack, called the intercept-resend attack, Eve intercepts the photon with a PBS oriented, for example, at 45°. Suppose that she finds a photon polarized at +45°. She does not know that the initial photon was vertically polarized, and she sends Bob a photon which is polarized at +45°. From the information shared with Alice, Bob knows that the photon should have been vertically polarized, but for Eve's photon, this holds true only in 50% of the cases. As Eve has a 50% probability of orienting her PBS in the right direction, Alice and Bob will register an error in 25% of the cases, and they will conclude that the photons have been intercepted by an adverse party. The safety of the protocol hinges on the fact that Eve cannot determine the photon polarization if she does not know in advance in which configuration it was sent.

Once the exchange of photons has been completed, Alice and Bob have both obtained a sequence of bits which are identical, at least in theory, the 001 sequence in the example of Figure 2.4. This sequence of bits is a random series of 0s and 1s and it will give them the key to be used later on to encrypt messages with a classical protocol and, of course, it must be known only to Alice and Bob. In practice, there are always errors due to detector and optical fiber imperfections and to possible attacks on the fiber. As explained above, Alice and Bob sacrifice a sub-ensemble of their sequence of bits and compare it publicly. This allows them to measure the Quantum Bit Error Rate (QBER), which is simply the probability that Bob measures the wrong value of the polarization when he knows that sent by Alice. Thanks to this QBER, they may now use a classical (that is, non-quantum!) error correcting code which allows them to obtain two strictly identical random series of bits. However, all these public exchanges may allow Eve to acquire some information on this series. Alice and Bob then use a further classical process, called privacy amplification, which allows them to establish a series of bits shorter than the initial one, but on which Eve does not have any information. This last series of bits will be used as a key. In the case of the BB84 protocol, it can be shown that the QBER must be less than 11% if one wants to establish a secure key. There is a final difficulty: in principle, Alice should send single photons, but due to the present limitations of single photon sources, she must in practice use attenuated laser pulses (Box 2.3). Furthermore, when one uses optical fibers, it is difficult to keep track of the polarization over long distances, and a different physical carrier for the bits, the phase of the photons, is used to implement the BB84 protocol. With this carrier, one can establish a key at a distance on order of 100 km with a rate of about 100 kbits/second.

Box 2.3. Attenuated laser pulses and single photons.

An attenuated laser pulse used in quantum cryptography typically contains an average of 0.1 photons. It is then possible to show that a non-empty pulse has a 5% probability of containing two photons (Appendix A.2.2), a fact which can be exploited by Eve to get some information on Alice's photons, but which can be countered by using a "decoy". If Alice sends single photons, the quantum no-cloning theorem guarantees that it impossible for Eve to deceive Bob although she can do better than a 25% error rate by using a more sophisticated strategy than the intercept-resend attack. For example, a strategy of partial cloning allows Eve to reduce to 16% the probability of sending Bob photons with an erroneous polarization state.

2.4.3 Limitations of quantum cryptography

The most serious limitation of quantum cryptography comes from the signal attenuation in the optical fiber. Typically, the intensity of a light pulse propagating in a fiber decreases by a factor of 100 over 100 km, and one must use repeaters which amplify the signal and allow it to recover its original shape and intensity. This is not possible for single photons which cannot be cloned, so that "photon amplification" is impossible. The maximum distance between two correspondents on a line for quantum cryptography is limited today to some 100 km, due to losses in the fiber and also to imperfections of the detectors. In particular, detectors have a non-negligible probability of triggering in the absence of any incident photon: this is called a dark count, which limits in a drastic fashion the signal-to-noise ratio of the communication. It is thus necessary to improve the transmission of optical fibers and/or to decrease the probability of dark counts. It is unlikely that the transmission in fibers will be improved in a significant way. On the other hand, there is steady progress in detectors, but that belongs to the domain of solid state physics (Chapter 6). Quantum physics brings a surprising solution, at least in theory, through the quantum teleportation protocol which was for a long time considered as a curiosity of fundamental physics, and which consists of transferring at a distance an unknown quantum state, for example, a photon polarization state. The success of such an operation is accompanied by an electrical signal able to trigger Bob's detector located at the end of the transmission line, conditioned by the success of the protocol. It is then possible to reduce the apparent noise in Bob's detector and to increase the signal-to-noise ratio, the price to be paid being a reduced transmission rate: this is the principle of a quantum relay. In order to obtain the quantum equivalent of repeaters, one should in addition use quantum memories which, however, are still in their infancy.

2.5 Further reading

The standard popularization book on cryptography is Singh [2000], which contains an introduction to quantum cryptography. Other good references at an intermediate level are Ekert [2006] and Sacarani *et al.* [2010]. At a more advanced level I would recommend Loepp and Wootters [2006]. Standard review articles on quantum cryptography are Gisin *et al.* [2002] and, on quantum cloning, Scarani *et al.* [2005].

3

Einstein, Bohr, and Bell

The final form of quantum physics, in the particular case of wave mechanics, was established in the years 1925–1927 by Heisenberg, Schrödinger, Born and others, but the synthesis was the work of Bohr who gave an epistemological interpretation of all the technicalities built up over those years; this interpretation will be examined briefly in Chapter 10. Although Einstein acknowledged the success of quantum mechanics in atomic, molecular and solid state physics, he disagreed deeply with Bohr's interpretation. For many years, he tried to find flaws in the formulation of quantum theory as it had been more or less accepted by a large majority of physicists, but his objections were brushed away by Bohr. However, in an article published in 1935 with Podolsky and Rosen, universally known under the acronym EPR, Einstein thought he had identified a difficulty in the by then standard interpretation. Bohr's obscure, and in part beyond the point, answer showed that Einstein had hit a sensitive target. Nevertheless, until 1964, the so-called Bohr–Einstein debate stayed uniquely on a philosophical level, and it was actually forgotten by most physicists, as the few of them aware of it thought it had no practical implication. In 1964, the Northern Irish physicist John Bell realized that the assumptions contained in the EPR article could be tested experimentally. These assumptions led to inequalities, the Bell inequalities, which were in contradiction with quantum mechanical predictions: as we shall see later on, it is extremely likely that the assumptions of the EPR article are not consistent with experiment, which, on the contrary, vindicates the predictions of quantum physics. In Section 3.2, the origin of Bell's inequalities will be explained with an intuitive example, then they will be compared with the predictions of quantum theory in Section 3.3, and finally their experimental status will be reviewed in Section 3.4.

The debate between Bohr and Einstein goes much beyond a simple controversy, which is after all almost eighty years old and has been settled today. In fact, the concept introduced in this debate, that of entanglement, lies at the heart of many very important developments of modern quantum physics, in particular all those linked to quantum information (Chapter 8). Moreover, we shall see that the phenomenon of non-local correlations compels us to revise in depth our space-time representation of quantum processes. These are the two reasons why a whole chapter is devoted to this debate.

3.1 Superluminal communications?

We communicate everyday with velocities close, but strictly less than the speed of light in a vacuum, $c \simeq 3 \times 10^8$ m/s, by exchanging light signals in the air, or through optical fibers, or more generally, by exchanging electromagnetic signals such as radio waves. It is well-known that special relativity forbids the exchange of information at a speed greater than c: in other words, superluminal communications do not exist. In special relativity, an *event* is defined by its location in space and time, the position where and the time when it occurs. For example the event "a car crash" is defined by its position $\vec{r}$ on the road and the time t when it happens. Two events may have a *causal relation* if it is possible that information travels from one event to the other at a speed less than, or equal to, the speed of light. In the opposite case, there cannot exist any mutual influence of one of the events on the other. This is called the *non-signaling* principle. However, special relativity does not forbid *displacements* at a speed greater than that of light. A simple example is given by the displacement at the surface of the Moon of the spot produced by a laser on Earth, whose diameter may be as small as a few kilometers (Section 4.5). If the laser rotates with an angular velocity larger than 10 revolutions per second, an elementary calculation taking into account the distance between the Earth and the Moon shows that the spot sweeps the surface with a speed greater than c. But this displacement cannot be used to exchange information, unless this information is *pre-established*, as in the following case: the laser beam may take two colors, green or red, and it sweeps the Moon surface from Alice to Bob who are located 100 km apart. If the beam is red, Alice raises her left arm, and if it is green, she raises her right arm. When the spot arrives at Bob, he knows which arm was risen by Alice, and this information was transmitted at a speed greater than that of light. However, this method clearly does not allow Alice to transfer to Bob any information on which they had not previously agreed. In order to truly exchange messages, they must contact the person who operates the laser on Earth.

Another example, borrowed from astrophysics, is the apparent motion of a relativistic jet emitted by a galaxy as observed by a physicist on Earth. For such an observer, the image of the jet emitted by the galaxy M87 travels at a speed which is 6 times that of light. Finally, special relativity is not incompatible with the existence of particles traveling at a speed greater than c, which are called *tachyons*. If they exist, tachyons *always* travel at a speed greater than c; they cannot slow down to smaller velocities. One can show that tachyons cannot allow exchange of information, except in science-fiction books. Whatever their origin, superluminal communications would lead to the so-called grandfather paradox: provided there is no privileged reference frame, you could travel back in time and kill your grandfather before he had any child, thus jeopardizing your very existence.

The following example of pre-established information will be expanded in the next section, allowing us to prove an important inequality. A box contains two T-shirts, one green and the other red, each of them wrapped in an opaque bag. After having met at a common point noted S (S for *source*), Alice and Bob each take one of the bags. They leave in opposite directions and, without opening their bag, travel 1.5 billion km. Having reached their final destination, they open the bags and put on their T-shirt. If Alice observes that she is dressed in red, she knows that Bob is dressed in green, and vice-versa. Although communication between Alice and Bob would take more than 3 hours, Alice knows *instantly* the color of Bob's T-shirt. Of course, there is no mystery: a correlation between the two colors was introduced at the source, and the apparent superluminal communication is nothing other than the result of pre-established information. The reader may rightly think that this example is trivial, and even preposterous, but he or she must be patient: a slight complication will lead to an absolutely non-trivial result in the next section.

3.2 A remarkable inequality

We shall enrich the wardrobe of Alice and Bob by attributing to them not only T-shirts but also pants, see Figure 3.1. The T-shirts and pants are as above either green or red, T_A and P_A denote the colors of Alice's T-shirt and pants, T_B and P_B those of Bob. In order to write down conveniently the catalog of their clothing, they assign numerical values, $+1$ or -1 to the colors: for example, $T_A = +1$ means that Alice's T-shirt is green, $P_B = -1$ that Bob's pants are red. This choice is a pure convention; the choice $+35$ and -578 would lead to the same final conclusions, at the price of a much more cumbersome algebra. The important point is that the variables T_A, P_A, T_B, P_B are *dichotomic variables*: they can take only one of two values, $+1$ and -1, or green and red. Alice and Bob have

Figure 3.1. Alice and Bob dressed with a T-shirt and pants. The figure represents a case where $T_A = T_B$, $T_A \neq P_B$, $P_A = T_B$, and $P_A \neq P_B$.

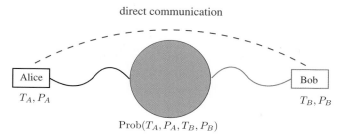

direct communication

Alice

T_A, P_A

Bob

T_B, P_B

$\mathrm{Prob}(T_A, P_A, T_B, P_B)$

Figure 3.2. The two possible origins of correlations between Alice's and Bob's stations. (a) Common causes: a (probabilistic) pre-established strategy of communication. This is the most general classical strategy of communication: the correlations arise from common causes. (b) Direct communication between the stations (green dashed line): Bob's results may influence those of Alice or vice-versa. If communication is limited by the speed of light, (b) is a particular case of (a).

4 possible ways of getting dressed, and there are in all $4 \times 4 = 16$ possibilities. The clothes are assigned according to a probability law where each of the 16 possibilities appears with a probability $\mathrm{Prob}(T_A, P_A, T_B, P_B)$, see Figure 3.2. The probability law allows us to compute, for example, $\mathrm{Prob}(T_A = T_B)$, the probability that Alice's T-shirt has the same color as Bob's, or $\mathrm{Prob}(P_A \neq T_B)$, the probability that Alice's pants and Bob's T-shirt have different colors. With these notations, the situation described in Section 3.1 corresponds to $\mathrm{Prob}(T_A = T_B) = 0$, $\mathrm{Prob}(T_A \neq T_B) = 1$. In the present example, a few lines of calculation show in full generality (Appendix A.3.1) that the following probabilities obey an inequality

$$
\boxed{
\begin{aligned}
&\mathrm{Prob}(T_A = T_B) + \mathrm{Prob}(T_A = P_B) + \mathrm{Prob}(P_A = T_B) \\
&+ \mathrm{Prob}(P_A \neq P_B) \leq 3.
\end{aligned}
}
\tag{3.1}
$$

Inequality (3.1) belongs to a quite general class of inequalities of probability theory discovered by Boole in 1862 and which Bell, unaware of Boole's work, re-derived one century later. Inequality (3.1), which is an example of *Bell's inequality*, is remarkable because all probabilities it depends on lie between 0 and 1, so the only obvious result is that the sum is less than 4, and yet the proof uses only elementary considerations. We have thus shown the following result: whatever the correlations introduced at the source by assigning clothes to Alice and Bob, inequality (3.1) is satisfied.

Box 3.1. Probabilities.

The probabilities $\mathrm{Prob}(T_A, P_A, T_B, P_B)$ obey the standard laws

$$0 \le \mathrm{Prob}(T_A, P_A, T_B, P_B) \le 1 \qquad \sum_{T_A, P_A, T_B, P_B} \mathrm{Prob}(T_A, T_B, P_A, P_B) = 1.$$

The probability $\mathrm{Prob}(T_A = T_B)$, for example, is obtained by summation

$$\mathrm{Prob}(T_A = T_B) = \sum_{T_A, T_B, P_A, P_B} \mathrm{Prob}(T_A = T_B, P_A, P_B).$$

3.3 Quantum physics and entanglement

Let us come back to quantum physics by giving a quantum version of the T-shirts and pants experiment. We shall use pairs of photons produced in a quantum state called *entangled* in polarization, a particular case of a general phenomenon: *quantum entanglement*. A source S, which can be built experimentally, delivers pairs of polarization-entangled photons whose instant of production is identified (Figure 3.3). Furthermore, the two photons of

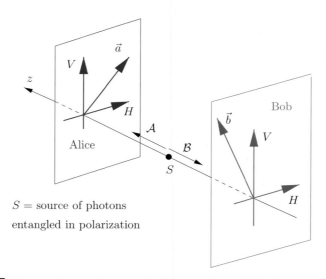

S = source of photons
entangled in polarization

Figure 3.3.   An experiment where a pair of photons in a polarization-entangled state is produced by a source S. The polarization of photon $\mathcal{A}$ is analyzed by Alice with an orientation $\vec{a}$ of her PBS, and that of photon $\mathcal{B}$ by Bob with an orientation $\vec{b}$ of his PBS.

the pair leaving the source travel in opposite directions along the z-axis toward the experimental stations of Alice and Bob, and their polarizations are correlated; if Alice measures the polarization of her photon $\mathcal{A}$ using the $\{VH\}$ configuration (see § 2.3.1) and finds a vertical (horizontal) polarization, then Bob, using the same configuration, will measure a vertical (horizontal) polarization of his photon $\mathcal{B}$. There is thus a perfect VV or HH correlation of the polarizations, each case occurring with a 50% probability. The results hold true whatever the configurations used by Alice and Bob for their polarization analysis, provided they use the same orientation of their PBS: for example, if they use the $\{DA\}$ configuration (§ 2.3.1), they will observe DD or AA correlations with 50% probability each. This kind of correlations, first discovered by EPR, can be reproduced easily by a classical probabilistic model. But the entangled state of quantum physics is not just a probabilistic mixture of states VV and HH, it is a *linear superposition* (§ 1.3.3) $(VV + HH)/\sqrt{2}$ (or $(DD + AA)/\sqrt{2}$) and, as we are going to see soon, *the correlations governed by this entangled state are much stronger than those of any classical probabilistic model*. The exact definition and mathematical description of entangled states may be found in Appendix A3.2.

Box 3.2. Proposal for superluminal communications.

One could think of exploiting these one-to-one correlations to establish superluminal communications between Alice and Bob in the following way. For the measurement of her photon, Alice uses either the $\{VH\}$ configuration, or the $\{DA\}$ configuration (§ 2.3.1). Just after Alice's measurement, Bob measures the polarization of his photon: if he finds V or H, then he knows that Alice used the first configuration, if he finds D or A, the second one. Bob knows immediately the configuration used by Alice, which could be used to exchange information at superluminal speeds. The idea is clever, but it does not work! The reason is that Bob cannot determine the unknown polarization state of his photon, although this information is known to Alice. The no-cloning theorem (§ 2.3.1) forbids this determination.

As already noticed, the correlations described previously are not at all surprising when Alice and Bob use the same configuration. Bell's breakthrough was to analyze a situation where Alice and Bob use *different* configurations to analyze their photons. The polarization of photon $\mathcal{A}$ is analyzed with a PBS whose axis is oriented along a direction $\vec{a}$ (of course, perpendicular to the direction of propagation) and that of $\mathcal{B}$ with a PBS whose axis is oriented along a direction $\vec{b}$ (Figure 3.3). Actually, a slight complication is needed: Alice may use two different axes $\vec{a}$ or $\vec{a}'$, Bob $\vec{b}$ or $\vec{b}'$, and the choice of axes is made while each photon is traveling from the source to the analyzers. The choice of axes is conventionally called the *setting* of the experiment. Alice and Bob are assumed to be located far enough from each other so that the orientation chosen by Bob cannot influence the results of Alice and vice-versa, a property called *setting independence*. It is also important that Alice and Bob may choose freely the orientation of their analyzers, or, in other words, they exert their *free will*. The polarization measurement, say, by Alice, gives two possible values: either the polarization is parallel to $\vec{a}$, which we write $A = +1$, or it is perpendicular to $\vec{a}$, which we write $A = -1$. If she uses the orientation $\vec{a}'$, we write the results as $A' = \pm 1$, and similarly Bob's results are written $B = \pm 1$ for orientation $\vec{b}$ and $B' = \pm 1$ for orientation $\vec{b}'$. We may now establish a one-to-one correspondence with the preceding section.

Alice's T-shirt $\qquad \Longleftrightarrow \qquad$ orientation $\vec{a}$

T-shirt color $T_A = \pm 1 \quad \Longleftrightarrow \qquad$ result $A = \pm 1$ of the polarization measurement

and more generally

$$T_A \Longleftrightarrow A, \quad P_A \Longleftrightarrow A', \quad T_B \Longleftrightarrow B, \quad P_B \Longleftrightarrow B'. \qquad (3.2)$$

At this point, it is useful to summarize the general context of a Bell-like experiment: each of the two far apart stations may choose two possible settings, or orientations, and each of the settings give two possible results, $+1$ or -1. We may now transpose at once Bell's inequality (3.1) which becomes in this context

$$\text{Prob}(A = B) + \text{Prob}(A = B') + \text{Prob}(A' = B) + \text{Prob}(A' \neq B') \leq 3. \qquad (3.3)$$

In the case of a photon pair in the entangled state $(VV + HH)/\sqrt{2}$ as defined previously, quantum theory allows us to compute the various probabilities in (3.3) as a function of the angles between the orientations $\vec{a}, \vec{a}', \vec{b}, \vec{b}'$, for example the angle θ between $\vec{a}$ and $\vec{b}$, as

$$\text{Prob}(A = B) = \frac{1}{2}(1 + \cos 2\theta). \qquad (3.4)$$

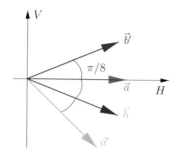

Figure 3.4. Configuration of the orientations $\vec{a}, \vec{a}', \vec{b}, \vec{b}'$ for a maximal violation of Bell's inequalities.

This equality is proved in Appendix A.3.2. If we choose the orientations of Figure 3.4, where the angles between $(\vec{a}, \vec{b})$, $(\vec{a}, \vec{b}')$, $(\vec{a}', \vec{b})$ are $\pi/8$ (22.5°) while that between $(\vec{a}', \vec{b}')$ is $3\pi/8$, and if we take into account $\cos \pi/4 = \sqrt{2}/2$ and $\cos 3\pi/4 = -\sqrt{2}/2$, we find

$$\text{Prob}(A = B) + \text{Prob}(A = B') + \text{Prob}(A' = B) + \text{Prob}(A' \neq B')$$
$$= 2 + \sqrt{2} \simeq 3.414 > 3. \tag{3.5}$$

Bell's inequality (3.1) is thus violated by quantum physics!

What went wrong with our derivation of inequality (3.1)? Yet the proof seemed quite general and the one-to-one correspondence between the wardrobes and the polarization measurements perfectly legitimate. The difference between the two situations is subtle, but fundamental: in order to evaluate, for example, the probability $\text{Prob}(T_A = P_B)$, Alice and Bob make a large number of trials where they look at the colors T_A and P_B of their T-shirt and pants respectively, and they may ignore the colors of their pants (Alice) and of their T-shirt (Bob), but the information on these colors exists, even though it is not used. In the experiment with entangled photons, when Alice chooses orientation $\vec{a}$ (she observes the color of her T-shirt), she cannot simultaneously measure along orientation $\vec{a}'$ *for the same photon*. Measurements along $\vec{a}$ and $\vec{a}'$ are *incompatible*; she can perform one or the other, but not both at the same time. If we want to pursue the analogy with T-shirts and pants, we have to admit that the value of the polarization along $\vec{a}'$ exists, even though Alice cannot measure it if she chose orientation $\vec{a}$. With such an assumption, Bell's inequality would be valid.

However, such an assumption is *at odds* with the principles of quantum mechanics: the choice of $\vec{a}$ is exclusive of that of $\vec{a}'$ and, as Asher Peres puts it, "unperformed measurements have no results". As we observed in § 1.3.3, the results of a measurement do not exist before they are observed and duly registered. Bell's inequality holds if the correlations are introduced at the source, as in the preceding section, through a probability distribution $\text{Prob}(T_A, P_A, T_B, P_B)$. The violation of this inequality by quantum physics shows that there is nothing like a probability distribution $\text{Prob}(A, A', B, B')$.

The notion of entanglement is one of the most important concepts of quantum physics. It is implicit in the EPR article, but at the same time, in 1935, it was Schrödinger who had the deepest understanding of it, and his account is still worth reading today.

> "When the two systems, of which we know the states by their respective representations (that is, their quantum states), enter into temporary physical interaction due to known forces between them, and after a time of mutual influence the systems separate again, they can no longer be described in the same way as before, viz., by endowing each of them with a representation of its own. I would not call that one but rather *the* characteristic trait of quantum mechanics, the one that enforces its entire departure from classical lines of thought. By the interaction, the two representations have become entangled."

An entangled pair of particles is such that each of its constituents loses its individuality, which is often called the lack of *separability* of the pair. Actually, entanglement is a *consequence* of the superposition principle, which allows us to build superpositions of *VV* and *HH* polarization states, in the case of photons entangled in polarization, but we may rightly say that entanglement is the most spectacular consequence of the superposition principle. To conclude this section, it is worth noting that it is possible to entangle two particles which have never interacted in the past, thanks to a procedure called entanglement swapping, which shares some similarity with teleportation.

3.4 Bell's non-locality and experiment

Up to now, we have only shown that quantum correlations are stronger than classical ones since they cannot be reproduced with a classical probability distribution, as witnessed by the violation of Bell's inequalities. Our next step will be to deduce the consequences of this violation for the space-time behavior of quantum processes. In classical physics, Galilean or Newtonian, the notion of causality is straightforward: an event can be influenced only by events which occurred at an earlier time, and it can only influence events occurring at a later time. In order for our arguments to be fully accurate, we need to couch our discussion in a relativistic framework, and then causality is defined thanks to the notion of *light-cone*. The light cone of an event O with space and time coordinates $t = \vec{r} = 0$ is the cone $c^2t^2 - \vec{r}^2 = 0$ (Figure 3.5a). The light cone of O may be divided into the past light cone, which contains all the events which could have influenced O, and its future light cone, which contains all the events that O might influence. These statements hold true because no signal can propagate at a speed greater than that of light. Two events with space-time coordinates $(t_1, \vec{r}_1)$ and $(t_2, \vec{r}_2)$ define a space-time interval

$$s^2 = c^2(t_2 - t_1)^2 - (\vec{r}_2 - \vec{r}_1)^2. \tag{3.6}$$

If $s^2 > 0$, we say that the two events are separated by a *time-like interval*, and if $s^2 < 0$, by a *space-like interval*. In the former case, the two events may be causally related, whereas in the latter there is no possible causal relation between them.

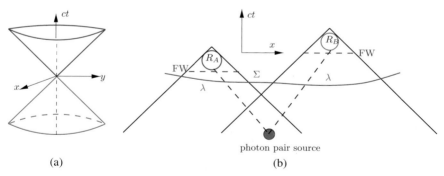

(a) (b)

Figure 3.5. (a) The light cone of an event O located at the origin of space-time coordinates drawn for one time and two space dimensions. The past light cone is shown in red and the future light cone in green. (b) The configuration of a Bell-type experiment drawn in two-dimensional space-time. The measurement of the polarizations for a given pair of photons takes place in space-time regions R_A and R_B, which are space-like separated. The decision for the choice of axis is taken at times defined by the two horizontal dashed lines FW (for free will). A space-like surface Σ intersects the past light cones of R_A and R_B.

Let us call R_A and R_B the two regions of space-time where Alice and Bob measure their respective photon polarization for one given pair emitted by the source (Figure 3.5b). These regions are assumed to be space-like separated: given any couple of events, one in R_A and the other in R_B, these events are separated by a space-like interval. The choice of axes $\vec{a}$ and $\vec{b}$, which we denote for simplicity by a and b from now on, is made just before the measurements begin, and this choice corresponds in Figure 3.5b to the lines FW (for free will). Let us call X the collection of all experimental devices (source, detectors, PBSs...) and λ a collection of parameters able to give a *complete theoretical* account of the experiment in any present or future theory. These parameters are assumed to be local, they are what Bell called "local beables", a counter example being given by an entangled state which is space and time independent, and spreads throughout space-time. If we draw a space-like surface Σ, that is, a surface such that any two points on it are space-like separated, intersecting the two past light cones of R_A and R_B, then, in order to determine the outcomes, we only need to know the theoretical parameters λ on the intersection of the surface Σ with the two past light-cones. Let $A = \pm 1$ and $B = \pm 1$ be the outcomes of the measurement, and let $\mathrm{Prob}(A, B|a, b, X, \lambda)$ be the probability of obtaining the results A and B conditioned on (a, b, X, λ). It is essential that the choice of axes (a, b) be entirely disconnected from the set (X, λ), so that they cannot be included in the common causes (Figure 3.2). They are chosen by Alice and Bob exerting their free will, or, more realistically, by an automatic device governed by a random generator attached to each station. As is demonstrated in Appendix A.3.1, the condition that outcomes in R_A and R_B depend uniquely on events located in their past light cones entails the factorization property

$$\mathrm{Prob}(A, B|a, b, X, \lambda) = \mathrm{Prob}(A|a, X, \lambda)\,\mathrm{Prob}(B|b, X, \lambda), \qquad (3.7)$$

where it is important to observe that the probability $\mathrm{Prob}(A|a, X, \lambda)$ does not depend on axis b, and conversely, the probability $\mathrm{Prob}(A|a, X, \lambda)$ does not depend on axis a. In full generality, λ is a random variable, and we must integrate over it with a weight $\rho(\lambda)$. As is shown in Appendix A.3.1, this integration leads to Bell's inequalities. Since Bell's inequalities are experimentally violated, as we shall see shortly, the conclusion suffers no ambiguity: the results in R_A and R_B cannot depend *only* on what happens in their respective past light cones, which contradicts the assumptions made by EPR. Although the terminology is not universal, this is what we shall call *Bell's non-locality*. Fortunately (or maybe not!), this property does not allow superluminal communication: *quantum mechanics is non-signaling*.

Let us formulate our results in a more intuitive language: correlations between results in R_A and R_B can arise only from common causes or from superluminal communication (see Figure 3.2). If we exclude the latter possibility, it remains only the former, that is, common causes, but this is also excluded if Bell's inequalities are experimentally violated. Assuming this to be the case, we see that there is no possible space-time picture for processes violating Bell's inequalities, which gives rise to *non-local correlations*. These non-local correlations do not allow signaling, so that the two concepts (Bell's) non-locality and non-signaling should be carefully distinguished. Be that as it may, *there seems to be no possible picture of non-local correlations arising from a continuous evolution in space-time*. In some sense, these correlations seem to emerge from "outside of space-time". The variables noted λ are often called "hidden variables", and one standard proof of Bell's inequalities relies precisely on their existence. We emphasize that in the preceding proof, which is inspired by Bell in his article "La nouvelle cuisine", the set λ is only assumed to give a complete theoretical description, in any present or future theory. Then the existence of hidden variables need not be assumed; it can be inferred from the preceding analysis. There exist many other proofs of Bell's inequalities, based on "realism", "local realism", "local hidden variables", "counterfactualness" etc., but none of them is as physically relevant as Bell's.

In the end, the decision must be left to experiment: do experimental results obey Bell's inequality (3.1) or, on the contrary, the quantum mechanical prediction (3.5)? It is important that the experimental conditions come as close as possible to the scheme drawn in Figure 3.5: the two measurement regions must be space-like separated, which is called the locality condition. Otherwise, the only thing we check is that quantum correlations are stronger than classical ones. Beautiful experiments were performed in the 1970s, but we had to wait until 1982 when Alain Aspect and collaborators were able to devise a two-photon source of unprecedented efficiency so that they could modify the orientation of the polarizers while the photons traveled between the source and the two stations. The distance between the stations was 12 m, traveled by light in 40 ns, and the analyzer orientation changed every 10 ns. Results confirmed without any doubt the violation of Bell's inequality and vindicated the prediction of quantum theory. A more recent experiment by Anton Zeilinger's group uses a still more efficient two-photon source, in which an ultraviolet photon is converted in a non-linear crystal into two photons with half of its energy, and which are polarization entangled (Figure 3.6). The orientation of the analyzers may be modified in a random way while the photons are flying from the source toward the detectors. This is possible, because the two stations are 400 m apart, a distance which is traveled by light in 1.3 μs, while the combined duration of individual measurements and rotation of analyzers takes less than 100 ns. It is then impossible that the choice of axes and the measurements at one station could influence those performed at the other one. In other words, this experiment satisfies the locality criterion. However, only 5% of the pairs are detected, and we have to assume, as in a poll, that this 5% sample is not biased. The opposite situation would be highly implausible but, given the importance of the result, this logical possibility should be addressed experimentally in order to close this so-called detection loophole. In order to do so, we would need to detect 80% of the pairs, instead of 5%. The detection loophole has been closed in an experiment made with ions, where the detection efficiency was close to 100%, but the two ions were only 1 m apart, so that the locality criterion could not be satisfied. In

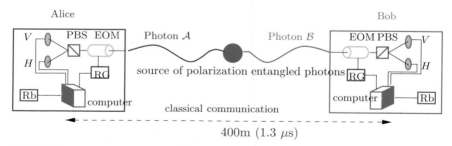

Figure 3.6.   The Innsbruck experiment by Anton Zeilinger and collaborators. Alice's and Bob's stations are 400 m apart, and the choice of the polarizer orientations is made by a random generator (RG) attached to each station, while the photons travel from the source to the stations. PBS = polarizing beamsplitter, Rb = rubidium atomic clock (Chapter 5), EOM = electro-optic modulator. Detectors are represented by yellow circles with red outlines. Experiment performed by Weihs *et al.* [1998].

the future, it is likely that an experiment will be able to close the detection and locality loopholes simultaneously. In the meantime, it seems most reasonable to assume that Bell's inequalities *are* violated.

3.5 Further reading

A good introductory book is Zeilinger [2010]. At a more advanced level the reader may consult Zeilinger [2006] and Maudlin [2010], who gives an overall review of the subject. Early experiments on the Bell inequalities were performed by Freedman and Clauser [1972], Clauser [1976] and Aspect *et al.* [1982]. The experiment of Anton Zeilinger's group is described in Weihs *et al.* [1998], and is reviewed by Aspect [1999]; a more recent experiment by the same group is Scheidl *et al.* [2010]. The experiment with entangled ions was performed by Matsukevitch *et al.* [2008]. Bell's paper "La nouvelle cuisine" is part of the collection of articles in Bell [2004], and Bell's argument is detailed in Norsen [2011]. The original Einstein/Bohr debate is found in the articles by Einstein *et al.* [1935] and Bohr [1935]. The case for superluminal communication is examined by Gisin [2013]. EPR correlations were used recently in an experiment proving non-invariance under time-reversal: Lees *et al.* [2012] and Schwarzschild [2012].

4

Atoms, Light, and Lasers

Up to now, the spatial properties of quantum particles played no more than a secondary role: we only needed the de Broglie relation (1.4) which gives the quantum particles wavelength, and our discussion of the quantum properties of photons was based mainly on their polarization, which is an *internal degree of freedom* of the photon. The probability amplitudes which we used did not involve the positions or velocities of the particles, which are *spatial,* or *external degrees of freedom*. In the present chapter, we shall introduce spatial dependence by defining probability amplitudes $a(\vec{r})$ that are functions of the position $\vec{r}$. In full generality, $a(\vec{r})$ is a complex number, but we shall avoid this complication and discuss only cases where the probability amplitudes may be taken real. For simplicity, we also limit ourselves to particles propagating along a straight line, which we take as the Ox axis: x will define the position of the particle and the corresponding probability amplitude will be a function of x, $a(x)$. In our discussion, we shall need to introduce the so-called potential well, where a particle travels back and forth between two points on the straight line. One important particular case is the infinite well, where the particle is confined between two infinitely high walls over which it cannot pass. This example is not at all academic, and we shall meet it again in Chapter 6 when explaining the design of a laser diode! Furthermore, it will allow us to introduce the notion of energy level, to write down the Heisenberg inequalities, to understand the interaction of a light wave with an atom and finally to explain schematically the principles of the laser.

4.1 Classical particles and classical waves on a line

4.1.1 One-dimensional potential well

Before we tackle the quantum case, let us recall some properties of a classical one-dimensional motion, by taking as an example that of the carriage of a roller coaster whose position is defined by its abscissa x on a horizontal axis. Figure 4.1 exhibits a possible motion of the carriage, where it starts by rolling on a horizontal part of the track. It arrives with some initial velocity at the downhill slope where it accelerates, reaches the bottom of the well and finally slows down on the uphill part of the track. In the absence of friction, and if the two horizontal parts of the track are at the same elevation, it will continue with its initial velocity. This example illustrates the physics of what is called a potential energy well or, in short, *a potential well*. Another interesting case which will be used later on is that where the carriage is released without initial velocity from some point A inside the well. Then, in the absence of friction, the carriage will reach uphill a point B located at the same elevation as A, and it will oscillate indefinitely between points A and B. When the amplitude of the oscillations is small, that is, when the elevation difference between A and the bottom of the well is small, we have an example of a *harmonic oscillator*, and the frequency f of the oscillations is fixed by the curvature of the track at the bottom of the well.

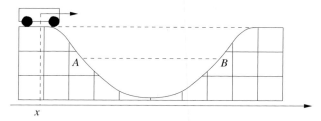

Figure 4.1. The roller coaster. The carriage comes from the left on a horizontal part of the track, it crosses the well and continues toward the right on the second horizontal part, at the same elevation as the left part. The carriage can also be released from point A without initial velocity and, in the absence of friction, it oscillates indefinitely between A and B.

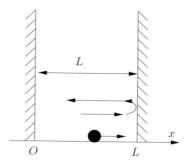

Figure 4.2. The one-dimensional box. The ball travels back and forth between the two vertical walls.

We shall replace the roller coaster carriage by a ball of tiny dimensions, a micro-billiard ball, the archetype of a classical particle, in order to facilitate the transition to the quantum case. A very simple example is that where the ball is confined between two infinitely high walls separated by a distance L (Figure 4.2), a case which we term *a particle in a one-dimensional box*, or a particle in a one-dimensional potential well with infinite walls. If the ball goes to the right with some velocity, it will be reflected by the wall on the right, and will start off toward the left with the same velocity. It will then be reflected by the wall on the left and will go back and forth indefinitely between the two walls in the absence of friction.

4.1.2 Classical waves

We just described the motion of a classical particle, but to go to the quantum case, we also need the wave aspect. What is the fate of a classical wave confined in a region $[0, L]$ and whose vibration amplitude $a(x)$ is a function of the position x between 0 and L? A good example is given by a string whose ends have abscissas 0 and L, and which is submitted at $x = 0$ to forced oscillations while the end $x = L$ is fixed (Figure 4.3: the x axis is vertical in this figure). This string possesses a large number of *vibration modes*, in theory an infinite number of modes. In the simplest vibration mode (Figure 4.3a), no point of the string (except, of course, the points $x = 0$ and $x = L$) stays motionless: we say that the vibration has no nodes. It will be convenient to label the number of nodes of a given mode by an integer n equal to the number of nodes plus one, that is, $n = 1$ in the case

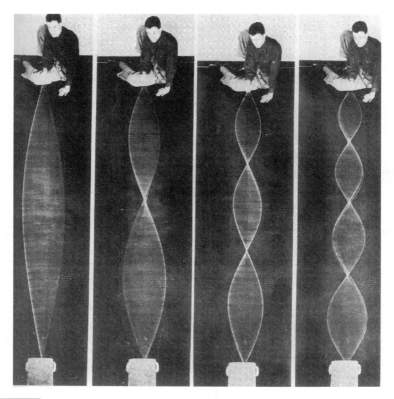

Figure 4.3. Stationary waves of a vibrating string. The x axis is vertical and the extremities of the string are fixed to the points $x = 0$ and $x = L$. From left to right, the vibration modes possesses zero ($n = 1$), one ($n = 2$), two ($n = 3$) and three nodes ($n = 4$) respectively. Reproduced from Hey and Walters [2003].

of zero nodes. This may seem a bit illogical, but the present convention turns out to be convenient in what follows. The next case is that where the vibration vanishes at the middle of the string, at $x = L/2$ (Figure 4.3b): the vibration has one node ($n = 2$), and the following modes have two nodes ($n = 3$, Figure 4.3c), three nodes ($n = 4$, Figure 4.3d), and so forth. The case $n = 1$ corresponds to the fundamental mode of vibration of the string, the case $n = 2$ to the first harmonic, the case $n = 3$ to the second harmonic etc. The preceding vibrations are called *stationary*, or *standing waves*, and they are characterized by a vibration amplitude which vanishes at $x = 0$ and $x = L$: $a(x = 0) = a(x = L) = 0$.

Instead of the standing waves in Figure 4.3, we can also observe *traveling*, or *progressive waves*, such as the waves at the surface of a pond described in § 1.1.1, which are obtained if the extremities of the string are not fixed. The standing waves in Figure 4.3 are in fact superpositions of waves traveling to the right and to the left (upward and backward in the figure), characterized by a wavelength λ, or a *wave vector* $k = 2\pi/\lambda$. In Figure 4.3(a), the wavelength is $\lambda = 2L$, and thus $k = \pi/L$. In the general case with $(n-1)$ nodes, $\lambda = 2L/n$, or, equivalently, $k = n\pi/L$. One shows in elementary physics that the vibration amplitude $a(x)$ is approximately a sine function, when it is small enough. For a mode with $(n-1)$ nodes, the amplitude $a(x)$ takes the following form

$$a(x) = a_n \sin kx = a_n \sin \frac{\pi n x}{L}, \quad n = 1, 2, 3, \ldots, \tag{4.1}$$

where a_n is a constant and $k = \pi n/L$. The function $\sin \delta$ is drawn in Figure 1.7; it vanishes at the origin and at $\delta = \pi$, $\delta = 2\pi, \ldots$, and one can check that $a(x)$ as defined in (4.1) does vanish at the origin and at $x = L$. The vibration frequency ν is approximately a linear function of the wave vector k, $\nu \simeq c_s k/2\pi$, where c_s is the sound velocity along the string. If the frequency of the fundamental mode is ν, that of the first harmonic will be approximately 2ν, that of the third 3ν and so on.

Box 4.1. Wave vector.

It may seem superfluous to introduce the wave vector in addition to the wavelength. However, in contrast to a wavelength which is always a positive number, the wave vector in dimension one is an algebraic number which can be positive or negative. In dimension three, it is a vector $\vec{k}$ related to momentum by $\vec{p} = h\vec{k}/2\pi = \hbar\vec{k}$, where $\hbar = h/2\pi$. The wavelength is proportional to the inverse of the modulus of the wave vector $|\vec{k}|$, $\lambda = 2\pi/|\vec{k}|$.

4.2 A quantum particle in a potential well

Let us substitute a quantum particle for the small ball of the preceding section and try to guess its behavior, given what we have learned about the particle and wave aspects of the motion restricted to the $[0, L]$ interval. As in any quantum description, we shall associate to the motion a probability amplitude, a complex number $a(x)$ depending on the particle position x, which we must interpret physically. From rule 2 of § 1.6.2, $|a(x)|^2$ is proportional to a probability, and it is natural to interpret it as being proportional to the probability of finding the particle at point x. More exactly, $|a(x)|^2$ is a probability density: $|a(x)|^2 \delta x$ is the probability of finding the particle in the interval $[x, x + \delta x]$, where δx is chosen small enough that the variation of $|a(x)|^2$ is negligible over this distance, a condition which reads in the present case $\delta x \ll L$. At this point we change our notations, in order to conform to the standard one: $a(x) \rightarrow \psi(x)$. The function $\psi(x)$ is called the *wave function* of the particle, and it is then the probability amplitude of finding it at point x.

Let us now turn to the determination of the wave function. The reasoning which we follow is not rigorous; it is what physicists call a heuristic one, but it allows us to uncover the underlying physics. The rigorous reasoning can be found in Appendix A.4.2. *In the absence of a potential well*, the particle would propagate along the Ox-axis, with a probability of finding it at any point on this axis which is independent of x, as all the point of the axis are physically equivalent. Furthermore, as the particle is a quantum one, it is endowed with a de Broglie wavelength λ. Therefore, in the absence of a potential well, the wavefunction $\psi(x)$ should obey the following two properties.

1. $|\psi(x)|^2$ is a constant which we may take equal to one, without loss of generality.
2. $\psi(x)$ is a periodic function of x with period λ.

It is worth observing (Box 4.2) that no real function obeys both conditions: the use of complex numbers is mandatory in quantum mechanics! In a classical wave point of view, the periodicity in λ gives rise to waves propagating along the string, and they can propagate either to the right or to the left. For a particle of mass m and velocity v, the wavelength λ is linked to the momentum $p = mv$ through $\lambda = h/p$, so that the wave vector is $k = 2\pi p/h$. In order to minimize the number of 2π's, one usually introduces the notation $\hbar = h/2\pi$. With this notation, the relation between wave vector and momentum simplifies to $p = \hbar k$, and it generalizes in dimension three to $\vec{p} = \hbar \vec{k}$ (Box 4.1).

Box 4.2. Progressive and stationary waves.

One may show that there exist only two functions which obey both conditions 1 and 2 as defined above: the complex exponentials $\exp(ikx)$ and $\exp(-ikx)$. One can indeed check *a posteriori* (see Appendix A.1.3) that, on the one hand, $|\exp(\pm ikx)|^2 = 1$ and that on the other hand, $\exp(\pm ikx)$ is periodic with period λ: as the wave vector is given by $k = 2\pi/\lambda$, and taking into account $\exp(\pm 2i\pi) = 1$

$$e^{\pm ik(x+\lambda)} = e^{\pm ikx}\, e^{\pm 2i\pi} = e^{\pm ikx}.$$

Furthermore, one shows that if $k > 0$, $\exp(ikx)$ represents a wave propagating to the right, and $\exp(-ikx)$ a wave propagating to the left. The function $\sin kx$, which represents a stationary wave, is a superposition of two waves traveling to the right and to the left respectively: $\sin kx = (e^{ikx} - e^{-ikx})/2i$.

Let us now return to the one-dimensional box of Figure 4.2: since the particle is confined in the $[0, L]$ interval, the probability of finding it outside this interval must vanish, so that $|\psi(x)|^2 = 0$ for $x < 0$ and $x > L$. If we demand that the wave function be continuous at $x = 0$ and $x = L$, this implies that the wave function must vanish at these two points. Furthermore, as the particle may travel to the right and to the left, $\psi(x)$ must be a superposition of de Broglie waves propagating to the right and to the left in the region $0 \le x \le L$. By analogy with (4.1), we can write $\psi(x)$ in the form

$$\psi(x) = a_n \sin kx = a_n \sin\left(\frac{\pi n x}{L}\right) \qquad n = 1, 2, 3, \ldots. \qquad (4.2)$$

The constant $a_n = \sqrt{2/L}$ is determined by demanding that the probability of finding the particle somewhere between $x = 0$ and $x = L$ be equal to one. From the relations $p = \hbar k$ and $k = \pi n/L$, we deduce that $p = \pm \hbar \pi n/L$, where the $\pm$ corresponds to the fact that the particle can travel to the right with a positive value $p = +|p|$ of the momentum or to the left with a negative value $p = -|p|$. The product $|p|L$ takes the value $\hbar \pi n$, and an important remark for the following section is that the minimum value of this product is reached for $n = 1$: $|p|L = \hbar \pi$.

4.3 Heisenberg inequalities and energy levels

The preceding discussion will allow us to introduce two fundamental properties of quantum physics: the *Heisenberg inequalities* (often

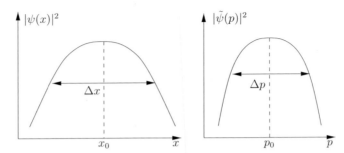

Figure 4.4. Schematic depiction of Heisenberg inequalities. In this example, the proba-
bility $|\psi(x)|^2$ exhibits a maximum at $x = x_0$, which gives the most probable value of the
particle position. The probability of measuring a momentum p is given by a function $|\tilde{\psi}(p)|^2$,
where $\tilde{\psi}(p)$ is the Fourier transform of $\psi(x)$. The most probable value of the momentum is
given by the maximum at $p = p_0$ of $|\tilde{\psi}(p)|^2$.

called Heisenberg's uncertainty principle, a terminology which should be
avoided), and the quantization of energy levels. Let us begin with the former. In quantum physics, a position measurement does not give a sharp
value, as the best available information is given by the *probability density*
$|\psi(x)|^2$. A single measurement gives a random value and, in order to obtain
information as precise as possible on the particle position, we must perform
a large number of measurements under identical conditions. The measure-
ment results will be spread according to the probability density $|\psi(x)|^2$,
as shown in Figure 4.4 (left), and the larger the number of measurements,
the better our knowledge of $|\psi(x)|^2$. The most probable value of the posi-
tion is given by the maximum of $|\psi(x)|^2$ at $x = x_0$, and the width of the
curve defines the dispersion Δx around the maximum. In the case of the one-
dimensional box, $\Delta x \sim L$, since the particle is localized with certainty in the
$[0, L]$ interval. Instead of measuring the particle position, we can measure
its velocity v or its momentum $p = mv$. In the case of the one-dimensional
box, we have seen that $p = +|p|$ or $p = -|p|$, depending on whether the
particle moves to the right or to the left, and the dispersion on the momen-
tum $\Delta p \sim |p| = \hbar\pi n/L$. The product $\Delta x \Delta p \sim (L) \times (\hbar\pi n/L) = \hbar\pi n$, and
its minimum value $\hbar\pi$ is reached for $n = 1$. The reasoning which we have
followed may be made rigorous: if we measure a large number of times the
position of a particle and its momentum in another series of experiments,
always under strictly identical conditions, then the product of the dispersion
Δx on the position and the dispersion Δp on the momentum are linked by

Heisenberg's inequality

$$\Delta x \Delta p \geq \frac{\hbar}{2}. \tag{4.3}$$

The dispersions Δx and Δp are in no way related to experimental uncertainties, and they depend only on the quantum state of the particle, as we have seen in the case of the one-dimensional box. The curves of Figure 4.4 are drawn for a generic quantum state. It must be emphasized once more that we consider repeated measurements performed on a particle which is always prepared in the *same* quantum state, and that position and momentum cannot be measured in the same experiment. Heisenberg's inequality implies that if a particle is prepared in a state in which its position is determined with a good accuracy, that is, if Δx is small, then a momentum measurement will give $\Delta p \sim \hbar/(2\Delta x)$, all the more large as Δx is small. This is the exact meaning of the often heard statement "it is impossible to assign simultaneously a position and a momentum to a particle". One should not draw from this statement the impression that quantum physics imposes fundamental limits on our knowledge: the role of Heisenberg's inequality is to set the limits which must be taken into account if we wish to stick to the classical concepts of position and momentum, even though we are dealing with situations governed by quantum physics.

Let us now turn to *energy quantization*. In the case of the one-dimensional box, the energy E of the quantum particle is purely kinetic, as the potential energy is x-independent and can be set to zero

$$E = \frac{1}{2} m v^2 = \frac{p^2}{2m},$$

where we have used $p = mv$ for the second expression of E. As the possible values of p are $p = \pm \hbar \pi n / L$, the energy takes *discrete* values E_n labeled by the integer n

$$E_n = \frac{\hbar^2 \pi^2}{2mL^2} n^2 \quad n = 1, 2, 3, \ldots . \tag{4.4}$$

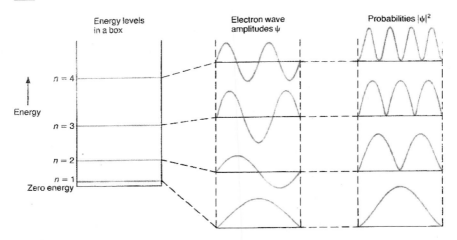

Figure 4.5.   The first four energy levels ($n = 1, 2, 3, 4$) and the corresponding wave functions of the one-dimensional box. Note that the probability of finding the particle at some point on the axis vanishes at the nodes of the wave function. Reproduced from Hey and Walters [2003].

Each value of E corresponds to an *energy level*. In classical physics, the energy of a particle in the box may take any positive value, and in particular the value $E = 0$, corresponding to a particle at rest and thus to $p = \Delta p = 0$, would be allowed. This situation is, however, incompatible with Heisenberg's inequality (4.3), as the product $\Delta p \, \Delta x$ would vanish. Another example is given by the potential well of Figure 4.1. The points A and B can be chosen arbitrarily, and then the carriage energy is equal to its kinetic energy when it goes through the bottom of the well. This energy is not quantized. Such a situation is forbidden in quantum physics, where only discrete values of the energy are permitted in the well. This property is called the quantization of energy levels. The state of lowest energy, E_1, is called the *ground state*. It plays an important role, because it is a stable state in the absence of external influences.

Energy level quantization lies at the basis of a fundamental property of atoms: an atom can jump from a higher energy level E_2 to a lower one E_1 by emitting a photon whose energy is, from energy conservation, equal to $(E_2 - E_1)$: $h\nu = E_2 - E_1$, hence the photon frequency ν. Conversely, if the atom is in the lower energy level E_1, and if it is shone with an electromagnetic wave of frequency ν, it will strongly absorb electromagnetic energy if $h\nu$ is close to the energy difference $(E_2 - E_1)$. If $h\nu = E_2 - E_1$, we say that

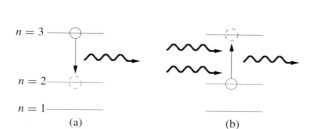

$n = 4$ ———— ————

$n = 3$ ————

$n = 2$ ————

$n = 1$ ————

(a)   (b)

Figure 4.6. Emission and absorption of photons by the one-dimensional box, taken as a model of an atom, between the levels $n = 3$ and $n = 2$. (a) Emission: the initial $n = 3$ state of the atom is represented by the green circle, the final state $n = 2$ by the red circle. (b) Absorption: two photons hit the atom whose initial state is $n = 2$ (red circle). A photon is absorbed and the atom final state is $n = 3$ (green circle).

there is a *resonance*, and it is at resonance that the rate of absorption of photons is the most important. The one-dimensional box is a crude model of an atom, but it is possible to check the properties we have just stated, because quantum wells rather close to our theoretical example can be experimentally built in semiconductors (Chapter 6). These quantum wells emit and absorb photons, and the frequencies which are emitted or absorbed are determined by energy levels: on account of (4.4), the possible frequencies v_{ln} are given by

$$hv_{ln} = E_l - E_n = \frac{\hbar^2 \pi^2}{2mL^2}(l^2 - n^2). \tag{4.5}$$

This equation expresses energy conservation for the emission of a photon (Figure 4.6). Indeed, if $E_l > E_n$ ($l > n$), the box emits a photon with energy $hv_{ln} = E_l - E_n$, and the initial energy E_l is equal to the final energy $E_n + hv_{ln}$. If, on the contrary, the box absorbs a photon, then $E_n > E_l$ ($n > l$), and the initial energy $E_n + hv_{ln}$ is equal to the final energy E_l.

4.4 Atoms

A frequently used picture of an atom is that of a miniature Solar System (Figure 4.7) where negatively charged electrons orbit a positively charged

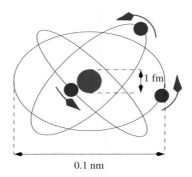

0.1 nm

Figure 4.7. The planetary model of an atom.

atomic nucleus, much smaller than the atom itself: the picture is reminiscent of planets orbiting the Sun. Remember that the dimensions of an atom are on the order of 0.1 nm (10^{-10} m), and those of an atomic nucleus on the order of a few fm (10^{-15} m): the atom radius and the nucleus radius differ roughly by a factor of 100 000. We shall choose for the sake of our discussion the simplest atom, the hydrogen atom, where the nucleus is a proton and a single electron orbits this proton. In the Solar System analogy, this would correspond to a single planet. The attractive gravitational force which maintains the planet on its orbit is replaced in the atom case by the attractive electrostatic (or Coulomb) force between the proton and the electron, and in both cases the force decreases according to an inverse square law $1/r^2$, where r is the distance between the electron and the proton, or between the Sun and the planet. This picture of the hydrogen atom is not entirely wrong, but it suffers a major flaw: in this picture, *the hydrogen atom is unstable*. Indeed, when an electron orbits a proton, it radiates electromagnetic energy because the motion is accelerated, and the atom behaves as a tiny antenna from which energy leaks into space. From energy conservation, the atom energy must decrease, and classical mechanics shows that the orbit radius must also decrease, so it spirals into the proton: the electron "falls" into the proton. However, we know that the atom radius, which can be defined in an approximate way, remains unchanged and on the order of 0.05 nm when the atom is in its state of lowest energy, or ground state, and the atom does not radiate, however long we wait. The stability of the atom contradicts the planetary picture.

Box 4.3 Energy of a circular orbit.

Elementary classical mechanics shows that the energy of a circular orbit with radius r is negative and proportional to $1/r$, $E = -C/r$, where C is a positive constant. The potential energy of the orbit is $-2C/r$, twice the total energy, and the kinetic energy is thus $K = C/r$. The value of the constant C is $e^2/2$, $e^2 = q^2/(4\pi\varepsilon_0)$, where q is the absolute value of the electron charge measured in Coulomb and ε_0 the vacuum permittivity. A more useful formula introduces the *fine structure constant* α, which is a dimensionless number given by $\alpha = q^2/(4\pi\varepsilon_0\hbar c) = e^2/(\hbar c) \simeq 1/137$.

Let us try to understand qualitatively why this catastrophic fall of the electron into the nucleus predicted by classical physics does not occur. *The key of the explanation is to be found in Heisenberg's inequality* (4.3). In our reasoning, we start from classical variables, the radius r and the absolute value p of the momentum on the orbit, as well as the classical expression of the energy. However, we are going to use Heisenberg's inequality to set limitations on the domain of validity of the classical approach: this is called a semi-classical reasoning. The Heisenberg inequality implies that the product rp is on the order of $\hbar$, $rp \sim \hbar$. If r becomes very small because the electron begins to fall into the proton, its momentum must become large, $p \sim \hbar/r$, and, as a result, its kinetic energy $p^2/2m$, where m is the electron mass, becomes very large. A general principle of physics states that a stable state corresponds to a minimum of the energy: thus, the stable state of the roller-coaster carriage of Section 4.1 is that where it is at rest at the bottom of the well (Figure 4.1), with a minimum potential energy. In our case, the potential energy of the electron is the electrostatic potential energy corresponding to the Coulomb force between the electron and the proton. This potential energy is negative and decreases to minus infinity when the radius tends to zero. When the electron gets closer and closer to the nucleus, its potential energy is large and negative, decreasing as $-1/r$, but the kinetic energy is large and positive, growing as $\hbar^2/(2mr^2)$, and thus faster than the absolute value of the potential energy. The stable state is obtained by looking for the best compromise between kinetic and potential energies, so that the total energy is as small as possible. A simple calculation (Appendix A.4.2) gives the radius and the energy of the ground state (see Box 4.3)

$$r = \frac{\hbar^2}{me^2} \simeq 0.053\,\text{nm}, \quad E = -\frac{me^4}{2\hbar^2} \simeq -13.6\,\text{eV}. \quad (4.6)$$

As a rule, it is Heisenberg's inequality which underpins the stability of atoms. As in the case of the quantum well, we have not only a ground state, but also excited states. The energy of the ground state is given by (4.6), and there exist an infinite number of excited states of the hydrogen atom, labeled by a positive integer $n = 1, 2, 3, \ldots$

$$E_n = -\frac{me^4}{2\hbar^2 n^2}. \tag{4.7}$$

As in the case of the quantum well, the frequencies of the photons which are absorbed or emitted by the hydrogen atom are given by $h\nu_{nl} = E_n - E_l$

$$h\nu_{nl} = -\frac{me^2}{2\hbar^2}\left(\frac{1}{n^2} - \frac{1}{l^2}\right). \tag{4.8}$$

The collection of energy levels of the hydrogen atom is called the *level spectrum* of this atom, from which its emission and absorption spectra follow. The case of an atom with many electrons is more complex, as there exists no exact formula such as (4.7) for the energy levels. However one still observes the phenomena of energy levels and energy spectrum. The emission and absorption spectra are characteristic of an atom, a kind of "fingerprint": two different atoms have different fingerprints. Some time ago, at the end of the XIXth century, astronomers discovered in the Sun an absorption spectrum different from any spectrum known on Earth at that time. The astronomers correctly deduced the existence of a new kind of atom, present in the Sun but not yet discovered on Earth: the helium atom.

4.5 Lasers

The interaction between matter (atoms) and radiation allows us to understand the principles of the laser, but we first need to give a more detailed picture of the interaction between an electromagnetic wave, say a light wave, and an atom. As we have seen, an atom in an excited state E_2 may spontaneously emit a photon and jump to a lower energy level, as in the case of Figure 4.6a. As we shall not need the other states, it is convenient to relabel $E_2 \to E_1$ and $E_3 \to E_2$. Conversely, the atom in its ground state E_1 may absorb a photon and jump to the excited level E_2. The *absorption* process is efficient only if the frequency of the incident wave multiplied by h

is close to the energy difference between the two levels: $h\nu \simeq E_2 - E_1$. As we have seen, this is the resonance condition.

Up to now, we have considered two possibilities: the *spontaneous emission* of a photon corresponding to $E_2 \rightarrow E_1$ and the absorption of a photon, $E_1 \rightarrow E_2$. As is implicit from its name, spontaneous emission occurs whatever the environment of the atom, but absorption can only exist in the presence of a light wave and the absorption rate is proportional to the light intensity. In 1916, in a seminal article, Einstein showed that there exists a third process, called *stimulated emission*. Assume that the atom in an excited state E_2 is shone with light of frequency ν such that $h\nu = E_2 - E_1$. The atom then emits a photon with a probability proportional to the light intensity: this process is precisely emission stimulated by the incident light. It is important to understand the difference between spontaneous and stimulated emission: in the former case, the emission is completely independent of the atom environment and the photon is emitted in a random direction; in the latter case, emission occurs only in the presence of a light wave, it is is proportional to its intensity (compare Figures 4.6a and 4.8), and the photon has the same characteristics as those of the incident light wave. In particular, it is emitted in the same direction.

A *laser* uses stimulated emission in order to amplify a light wave: LASER is an acronym for "Light Amplification by Stimulated Emission of Radiation". Let us consider the action of a light wave on a collection of N atoms in one of the two states E_1 or E_2. N_1 atoms are in state E_1 and N_2

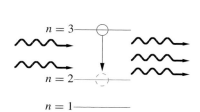

Figure 4.8. Stimulated emission between the levels $n = 3$ and $n = 2$. Two incident photons give rise to three photons in the final state, while the atom jumps from level $n = 3$ (green circle) to level $n = 2$ (red circle).

in state E_2 with, of course, $N_1 + N_2 = N$. If $N_1 > N_2$, the collection of atoms absorbs more energy than it delivers to the external world, since an atom in E_1 (E_2) absorbs (delivers) energy. In order to amplify the energy of the light wave, we must ensure $N_2 > N_1$. Unfortunately, because of Boltzmann's law (see Box 5.1), thermal equilibrium implies $N_1 > N_2$, while we wish to obtain $N_2 > N_1$, that is, *population inversion*. It can be shown that population inversion cannot be achieved with two levels only: we must use at least three levels, but the simplest and most efficient scheme, to be described later on, is a four-level system.

In general, a laser is composed of the following elements.

- An active medium which plays the role of an optical amplifier.
- An energy source which allows one to excite the active medium and to secure the population inversion. The energy source may be an electric one (see Chapter 6 on semiconductor lasers), or an optical one: a lamp or another laser. In this latter case, the process is called optical pumping.
- A resonant cavity which can be linear or ring-like (Figure 4.9), which selects the modes to be amplified. One of the cavity mirrors is partially reflecting, so that some radiation can exit the cavity. In fact, the laser converts pumping energy into light energy, or, more generally, electromagnetic energy. The presence of a resonant cavity is not compulsory, and there are other methods to select the modes to be amplified.

In the case of the ring cavity of Figure 4.9, the electric field must return to its initial value after a round trip, and the allowed wavelengths follow

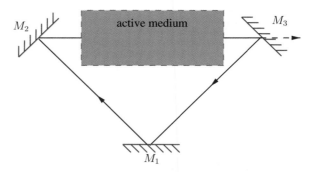

Figure 4.9.   Schematic design of a ring laser. The mirrors M_1 and M_2 are fully reflecting, while M_3 transmits part of the light, which will give rise to the laser beam.

from the condition that the phase be stationary: the optical length of the cavity, L, must be a multiple of the wavelength λ

$$L = n\lambda, \quad n = 1, 2, \ldots, \tag{4.9}$$

or, in terms of frequencies,

$$\nu_n = \frac{c}{L}n, \quad n = 1, 2, \ldots. \tag{4.10}$$

Of course, these conditions remind us of those giving the modes of the vibrating string in Section 4.2, with the difference that we now require periodic boundary conditions over a length L, rather than vanishing boundary conditions at the extremities of the string. It is often possible to select a single mode, and then to build a monomode laser. The laser beam has a finite transverse extension; its angular opening at large distances is governed by the laws of diffraction and is on the order of λ/d, where d is its minimum width, called the waist of the beam. This phenomenon is illustrated in Figure 6.8 in the case of a laser diode.

There exist several mechanisms which allow one to implement a population inversion of the levels E_1 and E_2. We shall limit ourselves to the simplest mechanism, which is based on four energy levels: E_0 (ground state), E_3 (excited state), and E_1 and E_2 (intermediate levels), see Figure 4.10. Optical pumping brings the atoms into level E_3 at a rate w. A fast relaxation then brings the atoms into level E_2, chosen in such a way that the relaxation from E_2 towards E_1 is slow; the relaxation time τ is the inverse of the rate $\tau = 1/\Gamma$. Finally a fast relaxation brings the atoms from E_1 to E_0. If we

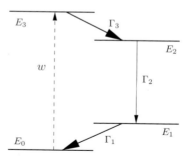

Figure 4.10. Sketch of a mechanism for population inversion with a four-level system. Optical pumping brings the system from its ground state E_0 to an excited state E_3 with a rate w. The thick arrows represent the fast transitions between levels.

call Γ_i the various relaxation rates, we must have

$$\Gamma_1, \Gamma_3 \gg \Gamma_2. \qquad (4.11)$$

It is easy to write down the equations which govern the time evolution of the populations N_0, N_1, N_2, N_3 of the four levels. For example, the population of level E_1 grows because of emission from level E_2 with a rate $\Gamma_2 N_2$, and it decreases because of emission toward level E_0 with a rate $\Gamma_1 N_1$. In a stationary regime, the population growth is exactly compensated by the population decrease, and thus

$$\Gamma_2 N_2 - \Gamma_1 N_1 = 0,$$

or, equivalently

$$\frac{N_2}{N_1} = \frac{\Gamma_1}{\Gamma_2}. \qquad (4.12)$$

In order to be fully accurate, we must add that we have supposed a non-saturated regime, where we can neglect stimulated emission or absorption between the levels E_2 and E_1. The full evolution equations are given, for example, in Grynberg *et al.* [2010]. The condition $\Gamma_1 \gg \Gamma_2$ does imply population inversion. Actually, stimulated emission is in competition with spontaneous emission, and the latter is detrimental to the operation of the laser. One shows that the rate of spontaneous emission is proportional to the cube of the frequency, ν^3, where ν is the frequency of the laser light. *The shorter the laser wavelength, the more harmful this emission is to laser operation.* Spontaneous emission is roughly eight times more important for blue light than for red light, and it is much easier to build a laser operating with red light than with blue light. Although the energetic efficiency is unfavorable, a green laser pointer is built from a laser diode emitting in the infrared, and the green beam is obtained thanks to frequency doubling in a non-linear medium.

The stationarity condition of the laser wave in the cavity is at the origin of the properties of time (or longitudinal) coherence and of spatial (or transverse) coherence of laser light, which underlie its remarkable behavior. One may, in principle, reach a coherence time which is limited only by spontaneous emission, and in practice coherence times on the order of 1 μs (and then coherence lengths of many kilometers) are easy to achieve. The spatial coherence allows the laser beam to approach the diffraction limit and to be focused on a spot whose dimensions are on the order of the wavelength. It is also spatial coherence which allows one to obtain a light beam whose divergence is close to one second of arc and, by dilating it to a diameter of 15 cm, to form on the Moon a spot of a few kilometers in diameter. Thanks to the reflection of the beam on a mirror, it is possible to measure the distance between the Earth and the Moon to a precision of a few millimeters.

Box 4.4. Longitudinal and transverse coherences.

If the light wave followed a perfect sine with frequency v, we would have at a fixed point in space a $\cos(2\pi v t - \phi)$ behavior. Actually, the phase ϕ is a (random) function of time, $\phi(t)$. If $\phi(t)$ has a given value at $t = 0$, the memory of this value will be lost after a time τ, the correlation time or coherence time, and $c\tau$ is the longitudinal coherence length. An analogous argument is used to define the coherence between two points in a plane perpendicular to the beam, which leads to the concept of transverse coherence.

If we compare a laser to a conventional light source, for example a frequency doubled YAG laser emitting green light with a power of 15 W, and a 150 W light bulb (an example borrowed from Grynberg *et al.* [2010]), the laser costs roughly $40 000 and the bulb $1: it looks like the laser represents a sophisticated and expensive technology for a relatively small power! But the laser takes a decisive edge when one attempts to concentrate the power onto a small surface or over a short time interval. Due to the laws of optics, it is impossible to obtain a light flux (a power by unit area) larger than 500 W/m^2 with the bulb. On the other hand, thanks to the spatial coherence of laser light, the power of the laser may be focused on an area whose size is limited only by diffraction, that is, by the wavelength $\lambda \sim 0.5\,\mu$m. The light flux may then reach 10^9 W/cm^2, a gain of more than one million with respect to the bulb. The time coherence allows one to concentrate the power over very short time intervals, with a high repetition rate, thanks to mode synchronization. For example, a titatnium–sapphire laser with synchronized modes allows one to reach a power of 10^{14} W over time intervals on the order of a femtosecond (10^{-15} s, femtosecond lasers).

One way of understanding the difference between laser light and conventional light is to compare the number of photons in a mode. In the case of sunlight, a light source corresponding to a temperature of 5500°C, the number of photons per mode in visible light is on the order of 1/100. By comparison, the number of photons per mode in a laser may be as large as 10^9. It is this large number of photons per mode which allows one to concentrate the power on small areas or over small time intervals.

There exist many different kinds of lasers: neodymium lasers (YAG), titanium–sapphire lasers, powerful CO_2 lasers . . . In everyday life, the most widespread lasers are the semiconductor lasers, or laser diodes, which are used for example to read bar codes, CDs and DVDs, and of course for optical fiber communications. Their principles will be explained in Chapter 6: the basic process is a recombination of an electron–hole pair, and the dimensions of the resonant cavity are on the order of a few hundreds of μm.

The laser is of course present in research laboratories, but also in the industrial and medical worlds. Important applications are grounded on the ability of lasers to supply energy in a concentrated form. A prominent advantage of the laser beam is that it is able to deliver energy to well-delimited parts of a material, without causing undesirable heating in their neighborhood. This property is used in surface treatments and in welding or cutting. The main medical applications are found in dermatology and ophthalmology. The coherence property of laser light also finds many applications: distance measurements, inertial navigation, remote sensing etc. Finally, inertial fusion could be a serious competitor to the magnetic one (ITER). The laser was invented just 50 years ago, but one could not imagine today a world without lasers!

4.6 Further reading

Further developments on atoms and lasers may be found in Hey and Walters [2003], Chapter 4. A useful reference at an intermediate level is Paul [2004]. Finally, lasers are treated at an advanced level by Grynberg *et al.* [2010].

5

Cold Atoms

This chapter and the following one address collective effects of quantum particles, that is, the effects which are observed when we put together a large number of identical particles, for example, electrons, helium-4 or rubidium-85 atoms. We shall see that quantum particles can be classified into two categories, bosons and fermions, whose collective behavior is radically different. Bosons have a tendency to pile up in the same quantum state, while fermions have a tendency to avoid each other. We say that bosons and fermions obey two different quantum statistics, the Bose–Einstein and the Fermi–Dirac statistics, respectively. Temperature is a collective effect, and in Section 5.1 we shall explain the concept of absolute temperature and its relation to the average kinetic energy of molecules. We shall describe in Section 5.2 how we can cool atoms down thanks to the Doppler effect, and explain how cold atoms can be used to improve the accuracy of atomic clocks by a factor of about 100. The effects of quantum statistics are prominent at low temperatures, and atom cooling will be used to obtain Bose–Einstein condensates at low enough temperatures, when the atoms are bosons.

5.1 What is temperature?

5.1.1   Absolute zero and absolute temperature

The notion of temperature is familiar in everyday life, when we use, for example, a thermometer built with a graduated tube linked to a reservoir containing a colored liquid. To calibrate the thermometer, we plunge it in a mixture of ice and water, and a mark is made on the tube corresponding to temperature zero. Then we plunge it in boiling water at standard pressure and a new mark corresponds to temperature 100. The interval between the two marks is then divided into 100 graduations, each corresponding to one Celsius degree (°C), and the graduation may be continued below zero and above 100. This definition of temperature is of course quite arbitrary, and another one is frequently used: temperature zero corresponds to that of a mixture of ice and salt in well-defined proportions, and temperature 100 to that of the human body. This second convention gives the Fahrenheit temperature (°F). Whatever convention is used, temperature defines an order relation: a thermometer allows us to decide that the temperature t_1 of an object is higher or lower than the temperature t_2 of a second object, $t_1 > t_2$ or $t_1 < t_2$. If the two objects are in contact, heat flows from the object with the higher temperature to that with the lower temperature. When $t_1 = t_2$, and if the two objects are in contact, there is no heat flow and we say that the two objects are in thermal equilibrium. However, the preceding definitions of temperature suffer from their arbitrariness, and moreover, the notion of an ordering does not allow an absolute comparison: we cannot say that an object is twice as hot as another one. In order to make quantitative comparisons, we must introduce the absolute temperature measured in kelvin (K).

Let us start with the following experiment: a dilute gas, nitrogen to be specific, is contained in a cylinder closed by a mobile piston (Figure 5.1). The whole system is put in a container in which the pressure is constant, but whose temperature, measured in °C, may vary. We begin with a high temperature, say 200°C, and decrease the temperature in intervals of, for example, 10°C. At each step, we wait long enough so that the gas is in thermal equilibrium with the container. Let us plot on a graph the volume

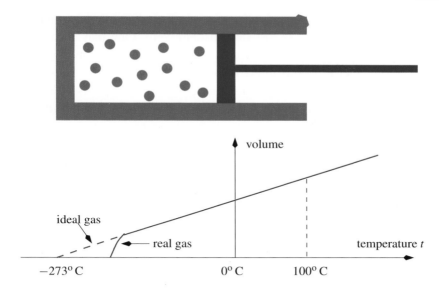

Figure 5.1. Experiment defining the absolute zero. At constant pressure, the volume of the gas depends on the temperature and decreases with it. The curve corresponding to a real gas (full line) stops before −273°C, because the gas becomes a liquid and its volume decreases suddenly.

of the gas as a function of temperature: we observe that the points lie on a straight line, and that this line intersects the temperature axis at $t = -273°$C. It is obvious that this point cannot be reached in the experiment, because nitrogen becomes liquid at a higher temperature, $t = -196°$C. Let us repeat the experiment with another dilute gas, oxygen, hydrogen, helium, xenon and so forth: as long as the gas is dilute and far from its liquefaction point, we recover the same straight line and the same intersecting point at $t = -273°$C, to which we may then attribute a universal value.

Physicists love idealizations, and in the present case, they will like to consider a gas of point-like molecules without interactions, which they call an *ideal gas*. In this case, there is no liquefaction, and the straight line may be continued down to $t = -273°$C. This point represents the lowest possible temperature, as for still lower temperatures, the volume would become negative. This universal temperature is called the *absolute zero*, and it is logical to adopt a new graduation of the thermometer, where zero temperature is just this absolute zero. One may then graduate the thermometer by deciding that the temperature of melting ice be $T = 273$ K, when the temperature is from now on measured in kelvins (K). With this new graduation, the temperature of boiling water is $T = 373$ K and room temperature about 300 K. The temperature T is the *absolute temperature*.

Several experimental groups have endeavored to obtain lower and lower temperatures. A temperature which is currently reached in order to cool the superconducting magnets of MRI (Magnetic Resonance Imaging) devices, is that of liquid helium, about 4 K. In the Large Hadron Collider (LHC) at CERN in Geneva (Chapter 7), 120 tons of helium are cooled down at a temperature of 1.9 K. In order to reach still lower temperatures, one uses a mixture of two helium isotopes, helium-3 and helium-4, which allows one to go down to 0.001 K, or 1 mK.

5.1.2 Absolute temperature and kinetic energy

A microscopic interpretation is able to give us an intuitive picture of temperature. In the example of dilute nitrogen enclosed in a container, the nitrogen molecules are not at rest: they move with random velocities on the order of 300 m/s at room temperature. If the container is motionless, the global momentum must vanish, and there are as many molecules flying in one direction as in the opposite one: the average velocity $\langle v \rangle$ vanishes. On the contrary, the average velocity *squared* $\langle v^2 \rangle$ is non zero, and statistical mechanics allows us to relate this average to the absolute temperature T of the gas for one component of the velocity, thanks to

$$\frac{1}{2} m \langle v^2 \rangle = \frac{1}{2} k_B T, \tag{5.1}$$

where m is the mass of the molecule and k_B the *Boltzmann constant*, whose numerical value is 1.38×10^{-23} J/K (joule per kelvin); $k_B T$ is a measure of the average kinetic energy $m \langle v^2 \rangle / 2$ of a molecule, often called energy of thermal motion. It is worth mentioning that (5.1) remains valid when the gas is not dilute, and even for a liquid, provided quantum effects are negligible. Equation (5.1) shows that the lighter the molecule, the larger its thermal velocity. At room temperature (about 300 K), this velocity is on the order of 1000 m/s for hydrogen and 300 m/s for oxygen.

We shall need for later purposes a slightly more refined description of thermal motion. As the molecules display random velocities, the detailed description is given by a velocity distribution function. For simplicity, we limit the discussion to the one-dimensional case, assuming that the molecules move on a straight line with a velocity v. The probability of finding this velocity in the interval $[v, v + \delta v]$, with a small enough δv, is by definition $\text{Prob}(v)\delta v$. The function $\text{Prob}(v)$ is drawn on Figure 5.2 for three different temperatures, $T_1 < T_2 < T_3$. The peak of the curve is centered at $v = 0$: as we have seen, the average velocity is zero. The width of the curve at mid-height, the dispersion $\Delta v = \sqrt{\langle v^2 \rangle}$, where $\langle v^2 \rangle$ is given by (5.1), is a measure of the velocity fluctuations around the zero average value. We observe that the width of the peak, Δv, decreases with the temperature. If we attempt to cool the gas down, the efficiency of the cooling will be measured by the width of the peak around $v = 0$. One should be aware of the limitations of the relation (5.1) between the absolute temperature and the average kinetic energy. This relation is valid only if quantum effects are negligible, and thus, as we shall see soon, if the temperature is not too low, but it is always valid for a classical fluid, whatever the density.

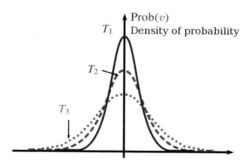

Figure 5.2. The velocity distribution function $\text{Prob}(v)$ for three different temperatures $T_1 < T_2 < T_3$. One observes the narrowing of the peak around $v = 0$ when the temperature decreases.

Box 5.1. Maxwell distribution and Boltzmann law.

When the average velocity $\langle v \rangle \neq 0$, then

$$\Delta v = \sqrt{(v - \langle v \rangle)^2}.$$

The velocity distribution function for $\langle v \rangle = 0$ has a Gaussian shape, called in that case the *Maxwell distribution* of velocities

$$\mathrm{Prob}(v) = \mathrm{const} \times \exp\left(-\frac{mv^2}{2k_{\mathrm{B}}T}\right).$$

In three dimensional space, the relation between kinetic energy and temperature is

$$\frac{1}{2}m\langle \vec{v}^{\,2} \rangle = \frac{3}{2}k_{\mathrm{B}}T.$$

$\langle \vec{v}^{\,2} \rangle$ is the average velocity squared, the average value of $\vec{v}^{\,2}$. We shall use (5.1), in order to avoid essentially irrelevant numerical factors. The three-dimensional generalization of Maxwell's distribution function is straightforward: if the velocity $\vec{v}$ has three components, (v_x, v_y, v_z), the velocity distribution function is a product of three identical functions, $\mathrm{Prob}(\vec{v}) = \mathrm{Prob}(v_x)\,\mathrm{Prob}(v_y)\,\mathrm{Prob}(v_z)$. The factor of 3/2 is easily understood: each space dimension brings a factor 1/2. Maxwell's distribution is a particular case of the *Boltzmann law*: the probability of observing a physical system in a state of energy E at temperature T is proportional to $\exp(-E/k_{\mathrm{B}}T)$.

5.2 Cooling atoms

5.2.1 Photon absorption and Doppler effect

In the preceding chapter, we have seen how an atom can jump from one energy level to another one under the influence of an electromagnetic wave with frequency v. It can, for example, jump from the ground state E_1 to an excited state E_2 by absorbing a photon of energy $hv = E_2 - E_1$. This absorption is accompanied by a recoil effect, which has been known for a long time and has been observed with conventional light. It is responsible of the radiation pressure which orients comet tails in a direction opposite to the Sun. However, it could only be quantitatively studied in laboratories after the advent of the laser, because the recoil effect is extremely small with conventional light. The recoil effect is familiar, and it is a consequence of momentum conservation. For example, in the cartoon of Figure 5.3, the

Figure 5.3. Because of momentum conservation, the unfortunate goalkeeper is thrown into the back of the net.

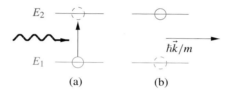

Figure 5.4. Atom recoil due to the absorption of a photon. (a) At first, the atom is at rest in its ground state. (b) It absorbs one photon with momentum $\hbar\vec{k}$ and, because of momentum conservation, it recoils with a velocity $\hbar\vec{k}/m$, where m is its mass.

unfortunate goalkeeper who attempts to stop the ball is thrown into the back of the net by a powerful kick: the initial momentum of the ball is transmitted to the goalkeeper, who recoils. Conversely, when a hunter fires a rifle, the initial total momentum is zero, and it remains so because the momentum of the bullet is equal and opposite to that given to the hunter, which is absorbed by his shoulder. At the very end, the hunter's momentum is absorbed by the ground. Momentum conservation is also at work when an atom absorbs a photon. Let us assume that the atom (the goalkeeper) is at rest and in its ground state E_1 (Figure 5.4). It absorbs a photon with a momentum $\vec{p} = \hbar\vec{k}$ and jumps into the excited state. We recall that $\vec{k}$ is the wave vector and $\hbar = h/2\pi$. Similar to the goalkeeper stopping the ball, the atom recoils and acquires a momentum $\vec{p}$, and thus a velocity $\vec{v} = \vec{p}/m$, called the *recoil velocity*. Although this recoil velocity is small, on the order of a few cm/s, one can easily imagine that repeating an absorption/emission cycle a large number of times leads to a fast slowing down of atoms emitted by an oven with velocities of a few hundred m/s, if a laser with a suitable frequency is shone on the atom beam. We shall see that the force on the atoms can be extremely large, up to 100 000 times the force of gravity.

5.2.2   Doppler cooling

In the first stage of laser cooling, atoms emitted by an oven with velocities of a few hundreds of m/s are slowed down by a laser, or a combination of lasers. We shall not give details on this first phase, which is somewhat complicated, and we turn directly to the second phase, where we shall trap atoms between laser beams. For simplicity, we shall assume that the atoms move along only one dimension of space, and in this case we need only two counter propagating laser beams. In the three-dimensional case, we would need six laser beams, two for each spatial direction. The physical phenomenon which is used in this trapping is the Doppler effect. Everyone has observed that the sound of an ambulance siren is raised in pitch when the ambulance approaches us and lowered when it travels away from us: this is the *Doppler effect* with sound waves. The frequency of the sound wave is first higher than that of the siren at rest when the ambulance gets closer, and then lower when the ambulance moves away. The same effect is observed with electromagnetic waves: the measurement of a car velocity with a radar gun (generally replaced today by a laser) uses the fact that the car reflects the radar signal, and it behaves as a source of electromagnetic waves which travel toward the receiver, the police officer. The larger the car velocity, the higher the frequency measured by the police officer will be. If the the radar signal is reflected by a car which moves away from the radar gun, the reflected frequency will be lower than that emitted by the radar gun. The dependence of the frequency on the velocity allows us to measure it, thanks to a quantitative formula for the Doppler effect: if a source moving towards an observer at rest emits an electromagnetic signal with frequency v, one can show that the frequency v' measured by the observer is given by (remember that for light $k = 2\pi v/c$)

$$v' = v(1 + v/c) = v + \frac{kv}{2\pi}. \tag{5.2}$$

Box 5.2. Radar guns and the Doppler effect.

The car moving towards the radar gun with a velocity v "sees" a frequency $v' = v + \Delta v$, where $\Delta v/v = v/c$. In the reference frame where it is at rest, the car emits a signal of frequency v', and the frequency measured by the receiver is $v' + \Delta v = v + 2\Delta v$. In the case of electromagnetic waves traveling in a vacuum, the frequency variation may depend only on the *relative velocity* of the source with respect to the observer. The exact formula derived from special relativity is, in the one-dimensional case

$$v' = v\sqrt{\frac{1 + v/c}{1 - v/c}}.$$

Taking into account $\sqrt{1 + \varepsilon} \simeq 1 + \varepsilon/2$ if $\varepsilon \ll 1$, we recover (5.2) if $v \ll c$.

Let us now return to atoms moving on a straight line. The two lasers emit in opposite directions with a frequency v, so that the configuration is that of two counter-propagating laser beams. The atoms are modeled by their two energy levels E_1 and E_2, with $E_2 - E_1 = hv_0$, v_0 being the resonance frequency (Figure 5.5). The laser frequency is chosen slightly lower than v_0, $v = v_0 - \Delta/2\pi$, where Δ is called the *detuning*. Let us recall that the absorption is maximum if $v = v_0$, that is, at resonance. If the atoms were at rest, they would absorb a relatively small number of photons, because the laser frequency is off resonance. But if the atoms have a positive velocity $v > 0$, they move toward the right hand laser and "see" a frequency $v' = v + kv/2\pi$, so that they are closer to resonance, they will absorb more photons than the atoms with a lower velocity and they will be slowed down because of the recoil. Conversely, the atoms moving to the left with a negative velocity $v < 0$ will absorb more strongly photons from the left hand laser, and they will also be slowed down. Whatever their

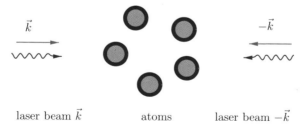

laser beam $\vec{k}$ atoms   laser beam $-\vec{k}$

Figure 5.5. Principle of one-dimensional Doppler cooling. The atoms in the cloud absorb photons from the two laser beams.

velocity direction, toward the right or toward the left, the atoms are slowed down. Note that in the three dimensional case, we would need six laser beams in order to slow down the atoms along the three spatial directions. We could be tempted to conclude that, after a suitable implementation of the process, the atoms could be slowed down to a vanishing velocity. But we have neglected a second process, which competes with absorption and the corresponding recoil, namely spontaneous emission. Indeed, when an atom initially in its excited state E_2 returns to its ground state E_1 by emitting a photon, this emission is also accompanied by a recoil (think of the recoil of the rifle when firing a bullet), so that the atom gains momentum. However, while the absorbed photons always have a fixed direction, that of the laser beams, the photons from spontaneous emission are emitted in a random direction. Under the various recoils, the atom experiences a random walk in velocity (or momentum) space. To sum up, there are two competing processes.

1. The absorption of photons along a direction which is fixed by that of the laser beams.
2. The emission of photons along random directions.

The absorption/emission process is called a *fluorescence cycle* (Figure 5.6).

The first effect, absorption, leads to a force opposite to the atom velocity, or momentum, typical of a viscous force of the form $\vec{F}_{\text{visc}} = -\gamma \vec{p}$, where γ is

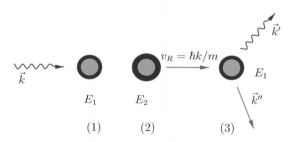

Figure 5.6. Fluorescence cycle. In phase (1), the atom at rest absorbs a photon from the laser with a fixed direction and it is brought into its excited state E_2. In phase (2), it is in its excited state with a velocity $\hbar \vec{k}/m$. In phase (3), it returns to its ground state E_1 while emitting a photon with wave vector $\vec{k}'$ in a random direction, where the atom momentum $\vec{k}''$ is given from momentum conservation by $\vec{k}' + \vec{k}'' = \vec{k}$.

the friction coefficient; this viscous force slows down the atoms. The second effect, spontaneous emission, leads to a random walk in momentum space, which means that the momentum increases proportionally to the square root of time. A stationary regime is obtained when the two effects balance each other. The atoms are not slowed down to a zero velocity but they keep some residual velocity. This residual velocity depends on a feature of the excited level which we have not yet introduced, its *mean life*, or *lifetime* τ. The mean life of an excited state is defined in the same way as that of a radioactive nucleus: if at time $t = 0$ we have a sample of N_0 nuclei, after a time $t = \tau = 1/\Gamma$ only half of the original number of nuclei, $N_0/2$ (more exactly N_0/e, $e \simeq 2.72$, see Box 5.3) will be present. The inverse Γ of the mean life is the *decay rate*. The same physics is at work for an atom in an excited state E_2: the atom does not stay indefinitely in this state, and if it is found in E_2 at time $t = 0$, the probability of finding it in the same state is only of roughly 50% at time $t = \tau = 1/\Gamma$. As in the preceding example, τ is the mean life of the excited state and for an atom Γ is called the *line width*. In the scheme of Figure 5.6, an atom undergoes absorption/emission cycles, and the rate of absorbed photons is controlled by τ, because, when an atom has absorbed a photon, we must wait on average a time τ in order to find it in its ground state, now ready to absorb another photon. In order to estimate the temperature which can be reached in the process, called the Doppler temperature T_D, we must look for an energy characteristic of the atom cycle, which is given by $\hbar\Gamma$. We may then identify the energies $k_B T_D$ and $\hbar\Gamma$, whence

$$k_B T_D \sim \hbar\Gamma. \qquad (5.3)$$

Box 5.3. Mean life and half life.

Let us examine the example of radioactive nuclei. The number $N(t)$ of nuclei at time t decreases as a function of the initial number of nuclei N_0 at time $t = 0$ according to an exponential law

$$N(t) = N_0 e^{-\Gamma t}. \qquad (5.4)$$

For a single nucleus, $\exp(-\Gamma t)$ represents the probability that this nucleus has not decayed at time t, also called its survival probability. The mean life $\tau = 1/\Gamma$ is the time when N_0/e nuclei have survived, and the half-life $\tau_{1/2} = \tau \ln 2$ the time when $N_0/2$ nuclei have survived.

Equation (5.3) is proved by demanding that the effects of absorption and emission balance each other in a stationary regime, and the full proof gives $k_B T_D = \hbar\Gamma/2$, which differs by a factor of 1/2 from our qualitative estimate, based on dimensional analysis. For the so-called D_2-line of sodium, for which $1/\Gamma = 1.6 \times 10^{-8}$ s, this temperature is $240\,\mu K$. The preceding discussion allows us to estimate the maximum acceleration of an atom in a laser beam. As the atom can suffer at most Γ cycles per second, and as the recoil velocity due to photon absorption is $\hbar k/m$, the maximal acceleration (in fact deceleration) is

$$a_{\max} \simeq \frac{\hbar k}{m}\,\Gamma.$$

For the D_2-line of sodium, this leads to $a_{\max} \simeq 10^5\,g$, 100 000 times the acceleration of gravity, $g \simeq 10\,\text{m/s}^2$. There exist cooling methods more sophisticated than Doppler cooling, which are based on the decomposition into sub-levels of levels E_1 and E_2, or sometimes of only one of them. These methods allow one to reach temperatures on the order of one microkelvin ($1\,\mu K$). *A priori*, it seems difficult to reach temperatures lower than the recoil temperature T_R whose origin lies in the recoil suffered by an atom when it is initially at rest and absorbs (or emits) one photon: as we have seen, the atom recoils with a velocity $\vec{v} = \hbar\vec{k}/m$, which corresponds to a kinetic energy $E_R = mv^2/2 = \hbar^2 k^2/m$ and to a temperature $T_R = E_R/k_B$. It looks like that one should not be able to go beyond this recoil limit, as every elementary phenomenon of the atom cycle corresponds to a variation E_R of the kinetic energy. However, by using subtle properties of quantum physics, researchers have been able to reach a few nanokelvins.

Up to now, we have examined atom trapping in velocity space: we have shown how to reduce the velocity to within a small velocity interval. In addition, it would be interesting to trap atoms in ordinary, or position, space, so that to confine atoms with low velocities into a volume of small dimension. This is made possible thanks to the *magneto-optical trap*, which acts as a restoring force on the atom: when the atoms move too far away from the center of the small volume, the restoring force takes them back toward this center. The magneto-optical trap uses the structure of energy levels, in the simplest case the excited one. The levels are split into sub-levels by a magnetic field, a phenomenon known as the Zeeman effect. A suitable

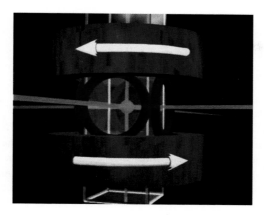

Figure 5.7. Artist's view of a magneto-optical trap. One can clearly see the six laser beams used for trapping and the cloud of trapped atoms. Published with permission from M. Le Bellac, *Quantum Physics*, Cambridge University Press, Cambridge (2006).

spatial variation of the magnetic field allows one to create a restoring force and to confine the atoms within a limited volume (Figure 5.7).

5.2.3   Atomic clocks and cold atoms

Cold atoms have allowed us to considerably improve the accuracy of atomic clocks. Clocks are always grounded on a periodic phenomenon, be it the oscillations of the pendulum of our grandmothers' clocks, or the vibrations of a quartz crystal of our present wristwatches. The periodic phenomenon used in atomic clocks is the vibration of the electromagnetic wave emitted in the transition between two well-specified levels E_1 and E_2 of an atom or an ion, characterized by its frequency ν_0

$$h\nu_0 = E_2 - E_1.$$

In order to build an atomic clock, we must be able to compare the frequency of the clock signal to that of the atomic transition. We must "interrogate" atoms, that is, compare the transition signal to that of the clock. This is illustrated in Figure 5.8 which compares the two signals. If we observe the two signals over a long enough time interval, we may be able to see a frequency difference, while an observation over a short time interval reveals no difference. The clock accuracy will be all the better if the duration T (not to be confused with the temperature!) of the comparison is long. The

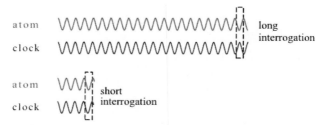

Comparison between the atom signal (red) and that of the clock (blue). The two vibrations coincide at time $t = 0$. If the signals are observed over a sufficiently long time interval, one clearly sees a difference (upper figure), while the signals are practically identical on a short time interval (lower figure).

comparison between the two signals allows us to correct for differences via a feedback mechanism and thus to lock the clock signal to the reference signal, that of the atomic transition. Atomic and ion clocks are at present the most accurate devices for the measurement of time. There exist many versions of these devices, one of the most widely used being the rubidium clock, which we find in GPS systems. This clock exploits a specific transition of the rubidium atom and is based on an interrogation method different from that of the Ramsey fringes, which we are going to describe soon in the case of the cæsium clock. It is of course necessary to eliminate as well as possible all side effects which could influence the transition. For example, the frequency of the transition is modified when the atoms are in motion, because of the Doppler effect.

As of today, the second is *defined* as the frequency ν_0 between two well-specified levels of the cæsium atom, which becomes then a universal standard of time. The frequency of the electromagnetic vibration emitted in this transition is *by definition* 9 192 631 770 Hz, about 10^{10} Hz. We shall explain one of the methods which are available to compare the frequency ν of the clock with that ν_0 of the atomic transition, the method of Ramsey fringes, whose principle is illustrated in Figure 5.9. An electromagnetic oscillator of frequency ν, the frequency of the clock that we wish to compare with that of the transition, feeds two cavities C_1 and C_2. The atom enters the cavity C_1, travels the distance between C_1 and C_2 over a time interval T, and crosses C_2. It is finally detected after exiting C_2 by the detector D, *provided it is in state E_2*. The principle of Ramsey fringes is analogous to that of a Mach–Zehnder interferometer. The cavities play the role of the

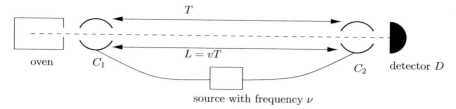

Figure 5.9. Principle of an atomic clock. The cavities C_1 and C_2 are fed by an electro-magnetic oscillator with frequency ν. The atoms arrive at C_1 in the level E_1, they travel from C_1 to C_2 during a time interval T, and thus over a distance $L = \upsilon T$, where υ is their velocity. They are finally detected in D if they are in the level E_2.

beamsplitters BS_1 and BS_2 of Figure 1.3. Let us call a_1 (a_2) the proba-bility amplitude of finding the atom in the level E_1 (E_2). When the atom enters C_1, it is in state E_1, the analogue of the propagation along the X-axis in the interferometer. The cavities act as beamsplitters: they transform the probability amplitudes according to

$$a_1 \rightarrow \frac{1}{\sqrt{2}}(a_1 - a_2), \quad a_2 \rightarrow \frac{1}{\sqrt{2}}(a_1 + a_2).$$

The origin of the minus sign in the first of the preceding equations is justified in Box 9.1. At the entrance of C_1, the amplitudes are $a_1 = 1$, $a_2 = 0$, as the atoms arrive in their ground state E_1. After crossing C_1, the amplitudes become $a_1 = a_2 = 1/\sqrt{2}$, from the preceding equations. After the atoms have crossed C_2, we have $a_1 = 0, a_2 = 1$. The probability of finding the atoms in the excited state E_2 is $\mathrm{Prob}(E_2) = a_2^2 = 1$, that for finding atoms in their ground state E_1 is zero. This is exactly the analogue of the Mach–Zehnder interferometer tuned in such a way that all photons are detected by D_Y. However, photons follow trajectories in ordinary space while, in the device in Figure 5.9, atoms follow the same spatial trajectories, but different "trajectories" in an abstract space, the space of quantum states. Starting from its ground state E_1, an atom is finally left after crossing C_2 in its excited state E_2. In a way, each cavity carries out half of the transition from E_1 to E_2. We must now take into account that our description is correct only at resonance, that is, $\nu = \nu_0$. Let us then assume that the frequency ν of the clock differs slightly from the resonance frequency ν_0: the detuning $\Delta = 2\pi(\nu - \nu_0) \neq 0$. This case is the exact analogue of the Mach–Zehnder interferometer, when the two arms are of different lengths (Figure 1.6).

One then shows that a phase shift $\delta = T\Delta$ is introduced between the amplitudes a_1 and a_2, and the probability of finding after C_2 the state E_2 is given by the analogue of (1.3)

$$\text{Prob}(E_2) = \frac{1}{2}[1 + \cos \delta] = \frac{1}{2}[1 + \cos (T\Delta)].$$

This effect allows us to compare the clock frequency to the resonance frequency, and, at the end of the day, to lock the frequency of the clock to that of the atomic transition. If the detector, which is sensitive to the excited state only, does not detect any atom, it means that $T\Delta = \pi$. We see that the method is able to detect a detuning $\Delta \sim 1/T$. T is the effective time for interrogating atoms, and the longer this interrogation time, the more accurate the clock will be. For standard atomic clocks using atoms exiting an oven, the relative accuracy is on the order of 10^{-13}, a drift of one second over 300 000 years. As the distance between the two cavities cannot be much larger than one meter, in order to increase T we must decrease the atom's velocity, which means that we must cool them.

An atomic fountain is a concrete realization of a clock using cold atoms (Figure 5.10). One first builds a cloud of cold atoms in an magneto-optical

Figure 5.10. An atomic fountain. The atoms trapped in the magneto-optical trap (horizontal arrows) are launched upward with a velocity on the order of 5 m/s. They cross the cavity (grey ring) twice, a first time on their upward motion and a second time on their downward motion, and they are detected by a laser (red beam). Copyright C. Salomon, Laboratoire Kastler-Brossel, Ecole Normale Supérieure; A. Clairon, SYRTE, Observatoire de Paris, CNES, Centre d'Etudes Spatiales. Courtesy of Christophe Salomon.

trap, which will later on cross a single cavity twice. The cloud is launched upward, and crosses the cavity a first time. After reaching an elevation of about one meter, it falls under the influence of gravity, and crosses the cavity a second time. The return trip time T, the effective interrogation time, is on the order of one second, instead of a few milliseconds in a conventional clock. This allows us a gain of a factor about 100, and the relative accuracy reaches 10^{-15}, a drift of one second over 30 000 000 years!

5.3 Bose–Einstein condensates

5.3.1 Quantum statistics and low temperatures

We saw that it is impossible to continue the straight line of Figure 5.1 down to zero temperature, because the gas becomes liquid before reaching the absolute zero and the dilute gas approximation breaks down. The transition from a gas to a liquid is the result of interactions between molecules, and one might think that if we considered a gas of point-like molecules without interactions, then the straight line of Figure 5.1 could be continued down to $T = 0$. Actually this is not true, and due to quantum effects, it is impossible to continue the straight line even for ideal gases. These quantum effects are linked to a property called *quantum statistics*. If the atoms, or molecules, of the ideal gas were micro-billiard balls, they could be distinguished by a thin line of paint, or a number. Painting a thin line on the balls or numbering them would modify in a negligible way their collision properties: when playing billiards, it matters little that the collisions involves two yellow balls, or two red balls, or one red and one yellow. Atoms are different in that it is impossible to paint them, or to number them, even in principle. Let us emphasize that this is not a practical impossibility, owing to our lack of a suitable technique, but it is an impossibility of *principle*. Two protons, two helium-4 or two rubidium-87 atoms are completely indistinguishable. On the contrary, two different isotopes of the same atom, for example rubidium-85 and rubidium-87 are distinguishable: the difference in nuclei is the "paint line" which allows us to distinguish them, even though their chemical properties are identical. Let us recall that two isotopes are atoms with the same number of protons and electrons, but a different number of neutrons. The notation rubidium-87, for example, means that the sum of the number of protons (37) and of neutrons (50) is 87. A helium-4 atom (2 protons and 2 neutrons) differs from a helium-3 atom, which has two protons but only one neutron, and a rubidium-87 atom differs from a rubidium-85 atom by two neutrons.

The indistinguishability of quantum particles suggests that nothing should change in their description when they are permuted. Let us begin with two particles which can be in quantum states i or j, and let $\psi(i,j)$ be the probability amplitude of finding the first particle in state i and the second one in state j. At first sight, we should recover the same probability amplitude after permuting i and j, since this is equivalent to a permutation of the two particles. However, this reasoning is not completely correct, because two amplitudes which differ by a phase factor (Box 5.4), for example a factor of -1, describe two physically equivalent situations: indeed, *probabilities* remain unchanged, and this is all we can demand, as only probabilities are directly observable, not amplitudes. Nothing prevents the probability amplitude after permutation from differing from the original one by a phase factor. It is quite remarkable that, according to a deep and rigorous theorem of relativistic quantum field theory, this factor can take only two values: $+1$ and -1: $\psi(i,j) = +\psi(j,i)$ or $\psi(i,j) = -\psi(j,i)$. When the phase factor takes the value $+1$, we say that the particles are *bosons*, and when it takes the value -1, we say that the particles are *fermions*. The bosonic or fermionic character of a particle is called its *statistics*. This transformation law under the exchange of two indistinguishable particles remains valid when we permute any two particles in a collection of identical particles. Electrons, protons, neutrons and neutrinos are fermions, and photons are bosons. A composite quantum particle made of an even number of fermions is a boson. Indeed, if we want to permute two such particles, we must permute an even number of fermions. To each permutation corresponds a (-1) factor, so that to a permutation of the two composite particles corresponds a factor $(-1)^N$, $(-1)^N = +1$ for N-even, and the particle is a boson. It is understood that the interactions of this composite particle do not break it down into its elementary, or more elementary, constituents. A similar reasoning shows that a particle composed of an odd number of fermions is a fermion. Thus, the helium-4 atom, composed of two protons, two neutrons and two electrons, is a boson, while the helium-3 atom, composed of two protons, one neutron and two electrons, is a fermion. The different statistics originate in the nuclei, the atomic nucleus is a boson in the former case, a fermion in the latter. Although the two isotopes of helium atoms have the same chemical properties and very similar behavior at ordinary temperatures — the slight differences may be linked to the mass difference between the two isotopes — their behaviors differ drastically at low temperatures, on the order of a few K, and this difference cannot be ascribed to the mass difference, but rather to the different statistics. Helium-4 becomes superfluid around 2 K, and helium-3 only at 2 mK! To summarize: quantum particles obey two types of statistics: bosons obey the Bose–Einstein statistics, and fermions obey the Fermi–Dirac statistics. As already mentioned, the existence of identical particles and of the two kinds of statistics has a deep origin in relativistic quantum field theory, which we shall come back to in Chapter 7.

Box 5.4. Phase factors.

A phase factor is a complex number of modulus one, and can be written in the form $\exp(i\theta)$. If $\psi(i,j)$ is the probability amplitude of finding two identical particles in states i and j, we can write *a priori*

$$\psi(i,j) = e^{i\theta}\,\psi(j,i).$$

It turns out that $e^{i\theta} = \pm 1$, $\psi(i,j) = +\psi(j,i)$ for bosons and $\psi(i,j) = -\psi(j,i)$ for fermions. This property is specific to the three-dimensional nature of ordinary space. In dimension two, it is possible to have particles which are neither bosons, nor fermions, but which belong to an intermediate class called anyons.

The example of helium shows that the effects of quantum statistics is most visible at low temperatures. As a general rule, at very low temperatures, *the kinetic energy is no longer proportional to the absolute temperature*, as in equation (5.1). Why is it so? In order to give a qualitative explanation, let us compare the de Broglie wavelength of particles confined in a trap with the average distance between them. As we saw in equation (5.1), the thermal kinetic energy of a particle is about $k_B T$, its velocity $\sqrt{k_B T/m}$ and its momentum $\sqrt{m k_B T}$. The corresponding de Broglie wavelength, called the *thermal wavelength*, is, according to (1.4), $\lambda_T \sim h/\sqrt{m k_B T}$. If n is the particle density, the average distance between two particles is $d \sim n^{-1/3}$. The minimum "quantum extension" of a particle is a wave

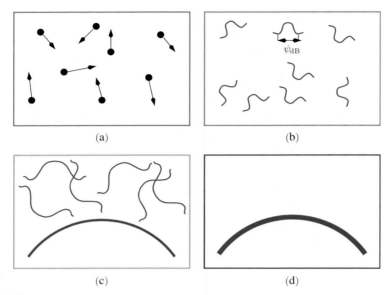

(a)   (b)

(c)   (d)

Figure 5.11. Variation of the de Broglie wavelength as a function of the temperature in the bosonic case. The temperature decreases from (a) to (d). In the upper left figure (a), the temperature is very high and the bosons can be viewed as micro-billiard balls. As the temperature decreases, the wavelength becomes on the order of the distance between the bosons (b), and then on the order of the dimensions of the trap (c). Finally, in (d), all the bosons are condensed in the ground state of the trap. The thick line represents the wave function of this state.

packet formed with de Broglie waves, and it is then on the order of λ_T. When $\lambda_T \ll d$, the particles keep their individuality. On the contrary, when λ_T becomes on the order of d, the wave packets begin to overlap, and the effects of quantum statistics become important (Figure 5.11). We thus expect important quantum effects when the temperature T becomes lower than a typical temperature T_c given by $d \sim \lambda_T$, so that

$$k_B T_c \sim \frac{h^2 n^{2/3}}{m}. \tag{5.5}$$

To summarize, for $T \gg T_c$, quantum effects are negligible, while for $T \ll T_c$, they are dominant. At that point, we must distinguish between bosons and fermions. The present chapter is devoted to bosons, the following one to fermions.

5.3.2 Atomic Bose–Einstein condensates

Let us consider a gas of bosons confined in a potential well. A single boson in the well has energy levels $\varepsilon_0, \varepsilon_1, \varepsilon_2, \ldots$, where ε_0 is the ground state energy. At zero temperature, the system tries to find a state of minimum energy, which corresponds to putting all the bosons in the ground state, ε_0. This is possible because there is no limitation to the number of bosons we can put in the same state, but we shall see that this strategy does not work for fermions. The bosons behave gregariously, they like to pile up in the same state, while fermions are staunch individualists, it is impossible to put more than one fermion in a quantum state.

At non-zero temperatures, the theory of the bosonic gas becomes complicated, even in the idealized case of point-like particles without interactions. It will only be possible to give a qualitative description, without entering a quantitative justification. To be specific, let us take the case of a collection of bosons confined in a magnetic trap which plays the role of a potential well with the shape of Figure 4.1. This well displays a roughly parabolic bottom and is called a harmonic trap. This means that the bosons are submitted to a restoring force $\vec{F}$ proportional to the vector $\vec{r}$ joining the center of the trap to the position of the boson: $\vec{F} = -m\Omega^2\vec{r}$, where $\Omega = 2\pi f$ is related to the trap frequency f, the frequency of the oscillations in the trap. This is analogous to the force exerted by a spring. To be fully correct, we must mention that the restoring force is not in general the same in all three directions, but we shall neglect this technical complication. In a harmonic trap, the energy levels are equidistant and labeled by an integer n: $\varepsilon_0 = 0, \varepsilon_1 = \hbar\Omega, \varepsilon_2 = 2\hbar\Omega, \ldots, \varepsilon_n = n\hbar\Omega, \ldots$ Let us assume that the trap contains $N \sim 10^6$ bosons, a typical value in present experiments. When the temperature is high enough, the probability of occupation of an energy level is given by Boltzmann's law (Box 5.1): it is proportional to $\exp(-\varepsilon_n/k_B T)$. When the temperature decreases and reaches a value lower than T_c, called the critical temperature, a finite fraction of the bosons is found in the level ε_0: this is the *Bose–Einstein condensation*. The critical temperature can be exactly calculated for a harmonic trap. One finds

$$k_B T_c = \frac{N^{1/3}}{1.06} \hbar\Omega. \qquad (5.6)$$

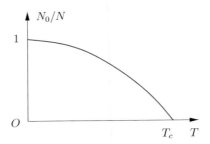

Figure 5.12. The ratio N_0/N giving the number of bosons in the ground state as a function of temperature. At $T = 0$, $N_0 = N$ and $N_0 = 0$ when $T > T_c$.

For $T < T_c$, the number N_0 of bosons in the ground state is given in Figure 5.12. When a finite fraction N_0/N of the atoms is in the ground state, they form a Bose–Einstein condensate.

Box 5.5. A few formulae for the condensates.

The numerical factor 1.06 in (5.6) is actually $(\zeta(3))^{1/3}$, where $\zeta(n)$ is the Riemann function familiar from number theory

$$\zeta(n) = \sum_{p=0}^{\infty} \frac{1}{p^n}, \quad \zeta(3) \simeq 1.202.$$

One shows that for $T < T_c$, the number N_0 of bosons in the ground state ε_0 for a gas without interactions, is given by

$$\frac{N_0}{N} = 1 - \left(\frac{T}{T_c}\right)^{3/2}.$$

In order to build a condensate, experimentalists begin with cooling down atoms in a magneto-optical trap, but this trap does not allow them to reach low enough temperatures. The atoms must be transferred into a magnetic trap and cooled down by "evaporative cooling" (Figure 5.13), a process which is analogous to that which allows us to cool down our coffee by blowing on its surface. The most energetic air molecules are located close to the hot coffee surface, and blowing on the surface allows us to eliminate them more rapidly than by waiting quietly for the coffee to cool down. The coffee surface then returns to a local thermal equilibrium with the air, but with a lower temperature since the most energetic molecules have been eliminated. In the atomic case, by a suitable tuning of the trap magnetic

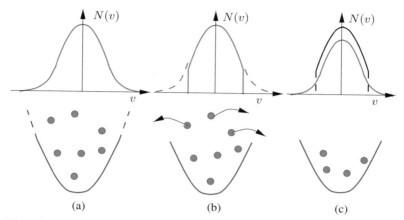

Figure 5.13. Evaporative cooling. The red curves give the number $N(v)$ of atoms with velocity v. In (a), the atoms are initially in thermal equilibrium, but the height of the trap (green) is suddenly lowered. In (b), the most energetic atoms, those with the largest velocities in absolute value, escape from the trap: the "wings" of the $N(v)$ distribution are chopped off. This new distribution is out of equilibrium and does not correspond to a well-defined temperature. In (c), the number of atoms in the trap is reduced, but the atoms have returned to thermal equilibrium, with a narrower distribution, and thus a lower temperature than in (a).

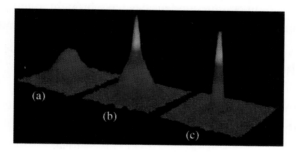

Figure 5.14. Bose–Einstein condensation. The figure displays the velocity distribution function of bosons. In (a), the temperature is above T_c by many μK, well above the condensation temperature $T_c \simeq 1 \, \mu$K, and the velocity distribution function is close to a Maxwell distribution (Box 5.1). In (b) the temperature is slightly above the critical temperature T_c. In (c) the atomic cloud is practically fully condensed: the velocity distribution is that of the ground state of the harmonic oscillator. Courtesy of W. Ketterle.

field, one manages to eliminate the fastest bosons and the remaining bosons return to thermal equilibrium, but at a temperature lower than the initial one. Repeating the process many times, one reaches a temperature low enough to form a condensate (Figure 5.14). One starts typically with 10^9 atoms in a magneto-optical trap in order to condense 10^6 atoms at a temperature around $1\,\mu$K.

All these experiments are extremely clever, but what is exactly their interest, in addition to confirming brilliantly Bose and Einstein's intuitions of 1921? As we have seen in the example of atomic clocks, low temperatures provide minute control of physical systems. Furthermore, a phenomenon called Feschbach resonances allows us to tune the strength of the interactions between bosons. This is the reason why the field of condensates witnessed a fast expansion during the past twenty years, giving rise to new ideas such as the study of the superfluid transition, trapping in optical lattices and extension to cold fermions. The low temperature control allows us to study models of condensed matter physics, such as the Hubbard model and other models useful for the study of high temperature superconductivity, which cannot be simulated on classical computers. The medium term perspective would be to develop analog quantum computers able to simulate quantum systems: as pointed out by Feynman, the computational power of classical computers required to describe an assembly of N quantum particles increases exponentially with N, which makes the problem intractable (Box 2.1). Low temperature condensates have brought together, in a spectacular and unexpected way, atomic physics and condensed matter physics!

5.4 Further reading

See Hey and Walters, Chapter 7; at an intermediate level, Cohen-Tannoudji and Dalibard [2006] and Phillips and Foot [2006]. The use of cold atoms in simulations is reviewed by Bloch [2013]. Finally, an overview of the subject is provided at an advanced level by Pethick and Smith [2001].

6

The Kingdom of Semiconductors

In this chapter, we move on to the case of fermions, and we shall find out that fermions are no less interesting than bosons! In practice, electrons are the most important example of fermions, because they are responsible for electrical conductivity in metals and semiconductors. It is impossible to understand a phenomenon as familiar as electrical conductivity without appealing to quantum physics. Two properties play a fundamental role: first the propagation of electron waves in crystal lattices, and second the Pauli exclusion principle, which is a consequence of the fermionic character of the electrons. In Section 6.1, we introduce electron wave propagation in crystals which gives rise to the phenomenon of energy bands, and we describe the filling of these bands according to the Pauli principle. These results will be used in Section 6.2 to describe the electronic properties of semiconductors, on which almost all our modern technology (laser diodes, optical fiber communication, computers, smartphones and so forth) is grounded. Finally, in Sections 6.3 and 6.4, we shall describe the principles of light emitting diodes (LEDs) and laser diodes.

6.1 Conductors and insulators

In the preceding chapter, we explained the tendency for bosons to pile up in the same quantum state. The situation for fermions is exactly opposite: it is impossible to put two (or more) identical fermions in the same quantum state. This property is termed *Pauli's exclusion principle*. This principle follows from the antisymmetrization of probability amplitudes in the case of fermions (§ 5.3.1). Let us consider two quantum states i and j occupied by two identical fermions, for example electrons, and let $\psi(i,j)$ be the probability amplitude of finding the first electron in state i and the second one in state j. As the two electrons are identical, we must have $\psi(i,j) = -\psi(j,i)$, while for two identical bosons we would have $\psi(i,j) = +\psi(j,i)$. If the two quantum states are identical, $i \equiv j$, then $\psi(i,i) = 0$, hence the Pauli principle: the probability amplitude of finding two identical fermions in the same quantum state vanishes. Atoms and molecules are built on this principle, because it imposes strong restrictions on the way energy levels are filled. We shall explain the general scheme with a particular example: filling the energy levels of a potential well. The level scheme is given in Figure 6.1, and the energy levels ε_n are the levels of single particles in the well. The fermions are confined in the well, and we look for the *global* ground state of a collection of N fermions, neglecting all possible interactions between them. This will be the ground state at zero temperature. We saw that for bosons this ground state is obtained by putting all the bosons in the lowest

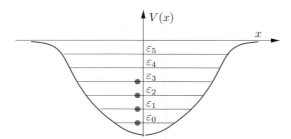

Figure 6.1. Filling the energy levels with fermions at zero temperature for a potential well $V(x)$. The levels are ordered from bottom to top starting with the ground state ε_0. In the example of the figure, we have to find room for 4 fermions, which occupy the levels $\varepsilon_0 \cdots \varepsilon_3$, while the levels ε_4 and ε_5 are not occupied. The Fermi level is ε_3 and the global energy $E_0 = \varepsilon_0 + \varepsilon_1 + \varepsilon_2 + \varepsilon_3$.

energy level ε_0, so that the global energy is $N\varepsilon_0$. This strategy does not work for fermions: once we have put one fermion in ε_0, we cannot put a second one in this state. In order to build the state with lowest energy, we must put the first fermion in ε_0, the second one in ε_1, the third one in $\varepsilon_3, \ldots$, until all fermions have found their place in one level. In other words, the levels are filled by order of increasing energy as shown in Figure 6.1. The highest energy level occupied by a fermion is of fundamental importance in solid state physics, and it is called the *Fermi level*. The Fermi level is ε_3 in Figure 6.1. Up to now, we have neglected interactions, so that our gas is an ideal gas of free particles. When we attempt to take into account interactions between fermions, the problem becomes extremely complicated, and has no exact solution. However, the notion of Fermi level remains valid: it is the energy which we must supply to the system in order to go from $N - 1$ to N fermions. In practice N is very large, $N \sim 10^{23}$, and it is also the energy necessary to add one particle, that is, to go from N to $N + 1$ fermions. The definition we have just given of the Fermi level is exactly that of the chemical potential in thermodynamics. The chemical potential and the Fermi level coincide at zero temperature, but this is no longer the case at finite temperatures. In spite of that, most semiconductor physicists use indifferently Fermi level and chemical potential.

We just described the case where N fermions move freely inside a potential well, for example, a three dimensional box, but in practice the most interesting case is that in which electrons move in the periodic potential created by ions in a crystal. A crystal is defined by a periodic structure, the crystal lattice, where the basic pattern, the *crystal cell*, is repeated periodically in space. A standard example is that of sodium (Na) chloride (Cl), or ordinary salt, which is represented in Figure 6.2. The Na^+ and Cl^- ions are found at the nodes of a periodic lattice, and the pattern is repeated in the three dimensions of space with a periodicity $a = 0.56$ nm called the *lattice spacing*. In a crystal, electrons close to the atomic nucleus which are strongly bound, or electrons of inner (atomic) shells, stay confined in the neighborhood of their respective nuclei and do not move in the crystal. On the contrary, external electrons, that is, electrons of outer shells, are weakly bound and may move more or less freely in the crystal. These are the electrons which will contribute to electric currents, because they are set into motion when submitted to an electric field resulting from an applied

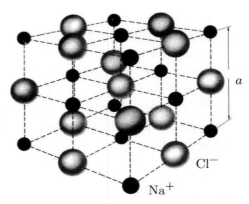

Figure 6.2. Crystalline structure of sodium chloride. The basic pattern is repeated along the three spatial directions, with a periodicity $a = 0.56$ nm, the lattice spacing.

voltage. At first sight, electrical conductivity in a metal seems to have an easy explanation, the motion of free electrons under the influence of an electric field. Electrons acquire some velocity v and, if the charge density is ρ, the current density is $j = \rho v$, which is directed opposite to the electron velocities due to the negative charge of the electrons. It is rather unfortunate that physicists at the end of the XIXth century chose to assign a minus sign to the electron charge: life would be much easier with a positive sign, as electric currents are mostly carried by electrons! In a more quantitative fashion, under the action of an electric field E during a time interval t, electrons acquire a velocity $-qEt/m$, where q and m are the absolute value of the charge and the mass of the electron. However, because of collisions, the electron velocity does not increase indefinitely with time but instead reaches, almost immediately, a limiting velocity proportional to E. This limiting velocity is actually that featured in the expression of the current density, and the microscopic equation $j = \rho v$ is then equivalent to the familiar Ohm's law. Electrical conductivity is an example of *transport process*, and there are many other examples, such as heat conduction or viscosity. Furthermore, the mass which appears in the acceleration is not the true electron mass as measured in a vacuum, but the effective mass m^* of an electron which propagates in the crystal lattice (Box 6.1).

Box 6.1. Quasi-particles and effective mass.

Actually, our assumption of freely moving electrons is far from warranted. It is quite remarkable, as was first pointed out by the Soviet physicist Lev Landau, that an electron gas with strong interactions can be approximately described as a gas of weakly interacting *quasi-particles*, provided we simply replace the mass m in a vacuum by the effective mass m^* of the quasi-particle. This effective mass may be measured in cyclotron resonance experiments. The quasi-particle scheme is far from universal. In systems with low dimensionality, it is not possible in general to ignore the interactions, and we then encounter systems of "strongly correlated fermions", whose properties are not at all similar to free particles and are not yet well understood. These strongly correlated systems seem to play an important role, for example, in high temperature superconductivity.

The mechanism for electrical conduction which we just described is indeed very simple, but it is completely unable to explain why there are conductors and insulators! In order to understand a phenomenon as simple as electrical conductivity, *we cannot bypass quantum physics*. When an electron moves in a crystal, we can in no way think of this electron as a micro-billiard ball; the wave aspect is essential. The propagation of electronic waves in the periodic potential created by the crystal ions is accompanied by multiple interference effects, and these interferences imply that the allowed energy levels are found only in well defined energy bands, called the *allowed bands*, while other energy intervals, the *forbidden bands*, are excluded (Figure 6.3a). The forbidden bands originate from destructive interferences, and they are typical wave phenomena. In a perfect crystal, at zero temperature, electrons propagate without any dissipation, or friction. The electrical resistance familiar from Ohm's law originates from lattice imperfections: impurities, defects or vibrations due to thermal motion of the ions. For simplicity, let us use once more the one-dimensional case. Because of the lattice periodicity, the wave vector $k + 2\pi/a$ is equivalent to k (see Box 6.2), where a is the lattice spacing, so that we can restrict k to the interval $0 \leq k \leq 2\pi/a$ or, in a more symmetrical way, to $-\pi/a \leq k \leq +\pi/a$, as in Figures 6.3 and 6.4. The first essential feature of quantum physics involved in the explanation of electrical conduction is the structure of energy levels in allowed and forbidden bands (Figure 6.3a). The second feature is the fermionic character of electrons. As in the example of the potential well, we must fill the levels in the energy bands with the free electrons of the crystal or, more exactly, the almost-free quasi-particles.

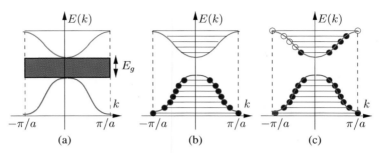

Figure 6.3. Energy bands and electrical conductivity. (a) Allowed and forbidden energy bands: the valence band (green) and the conduction band (red) are allowed bands. Each energy level $E(k)$ corresponds to two values of k, one positive and the other one negative. Between the allowed bands, there exist no energy level (grey zone): this zone is a forbidden band. The width of this forbidden band or, equivalently, the distance (in energy) between the maximum of the valence band and the minimum of the conduction band, is the band gap E_g. (b) Insulator or semiconductor: the valence band is full (full circles) and the conduction band is empty. The energy levels in the allowed band form a quasi-continuum. For clarity, the energy difference between successive energy levels of a single band has been greatly exaggerated. (c) Conductor: the conduction band is partially filled, the full (empty) circles correspond to the occupied (unoccupied) levels. Under the action of an electric field, the conduction band becomes asymmetric: there are, for example, more electrons with $k > 0$ (moving to the right) than electrons with $k < 0$ (moving to the left), which entails an electron flux directed from the left to the right, and then an electrical current in the opposite direction, owing to the negative charge of the electrons.

Two bands are going to play a crucial role, the *valence band* and the *conduction band*. The valence band corresponds to low lying levels, and it is completely filled (Figure 6.3b). The energy levels in the allowed bands form a quasi-continuum, and the energy difference between successive energy levels of a single band has been greatly exaggerated for clarity: the difference between two successive levels of a single band is typically 10^{-20} eV, while the energy difference between two bands is typically on the order of 1 eV, and these levels form a quasi-continuum. The valence band cannot contribute to electrical conductivity, because it is symmetric under the interchange of $+k$ and $-k$: to any electron with energy $E(k)$ in the valence band there corresponds another electron with the same energy $E(-k)$. To any electron of wave vector $k > 0$ and velocity $v = \hbar k/m$ oriented toward the right, there corresponds another electron with wave vector $-k$ and velocity $-\hbar k/m$, so that the global flux of charge vanishes. There is of course no current in the absence of an applied electric field, but the presence of such a field changes nothing: electrons stay blocked in their initial energy levels because of Pauli's principle. An insulator is a crystal where the valence band is full and the conduction band is empty (Figure 6.3b). On the contrary, the conduction band of a conductor is only partially filled. When an electric field is applied to a conductor, the $k > 0$ electrons, for example, may occupy higher energy levels (Figure 6.3c), by using levels which were previously unoccupied. This is forbidden for the valence band, because of Pauli's principle. To summarize, electrical conduction is grounded on two phenomena, which are typically quantum.

1. The existence of allowed and forbidden energy bands, which originate in the wave behavior of electrons in a periodic potential.
2. The filling of these bands in conformity with Pauli's principle.

Box 6.2. Momentum and wave vector in a crystal.

One can show, thanks to a theorem due to Bloch which is the equivalent of Floquet's theorem for time-dependent functions, that the wave function takes the form $\exp[ikx]u(x)$, where $u(x)$ is periodic with period a. If we substitute $k' = k + 2\pi/a$ for k

$$\exp[ik'x]\,u(x) = \exp[ikx]\,[\exp[(2i\pi x/a)]\,u(x) = \exp[ikx]\,u'(x)$$

where $u'(x)$ is also periodic with period a, so that we may restrict k to the interval $-\pi/a \le k \le \pi/a$. The wave vector k does not vary between $-\infty$ and $+\infty$ as is the case for a free particle, and k is called a "quasi-wave vector". Similarly, the momentum associated to k, $p = \hbar k$, is a "quasi-momentum", and not the momentum of a free particle.

We saw in the preceding chapter that we must reach very low temperatures to observe quantum effects: 2 K for helium-4 superfluidity, 1 μK for Bose–Einstein condensates. These are indeed very low temperatures compared to room temperature, about 300 K. In the case of the electron gas, we may associate to the Fermi energy ε_F a Fermi temperature $T_F = \varepsilon_F/k_B$. The Fermi energy of a typical metal is a few eV, so that $T_F \sim 10^4$ K. The Fermi temperature T_F is very large compared with room temperature. A conductor is a quantum gas for which room temperature is relatively low, and this explains why the electron gas within a conductor at room temperature behaves as a quantum, and not a classical, gas. One important consequence is that the average kinetic energy of the electrons is not proportional to the temperature: equation (5.1) is not applicable.

6.2 Semiconductors

Let us examine once more Figure 6.3b: when the temperature is non-zero, $T \neq 0$, thermal motion sends some electrons of the valence band into the conduction band. The electron density in the conduction band may be deduced from Boltzmann's law (Box 5.1) and is found to be proportional to $\exp[-E_g/(2k_B T)]$, where E_g is the energy necessary to jump from the maximum of the valence band to the minimum of the conduction band, called the *band gap*, or simply the *gap*, hence the notation. As E_g is on the order of one eV and $k_B T \sim 0.0025$ eV at room temperature, this electron density is very low. When the gap is small, $E_g \lesssim 1$ eV, it is however not entirely negligible and the crystal allows a weak current to flow under an applied electric field, which justifies the name *semiconductor*. The conductivity of thermal origin decreases as the gap increases, and in the case of carbon in the form of diamond, with a gap $E_g \simeq 5.5$ eV, it is practically zero: diamond is an excellent insulator. This being said, we see that there is no difference in principle between an insulator and a semiconductor.

Were thermal motion the only source of electrical conductivity in semiconductors, they would have ended in the dustbin of history. Semiconductors become interesting because we know how to dope them in order to boost their conductivity. One can, for example, start from a silicon crystal and then replace some of the silicon atoms with phosphorus atoms at some nodes of the crystal lattice. Phosphorus has one valence electron more than silicon and this additional electron is going to behave as a free electron in the lattice: the phosphorus is a *donor* (of electrons). Because of Pauli's principle, this electron must occupy a level of the conduction band, as all the energy levels of the valence band are occupied. We have then built a *n*-type semiconductor, where *n* stands for "negative", as the current carriers are negative charges, those of the electrons. This case is represented in Figure 6.4a.

The case represented in Figure 6.4b is also remarkable and is going to play a major role in semiconductors. In this figure, we observe that the conduction band is empty, but that a few electrons are missing in the valence band. One can show that the absence of an electron with a wave vector k is equivalent to an effective particle with a wave vector $-k$ and a charge opposite to that of the electron. This effective particle which, of course,

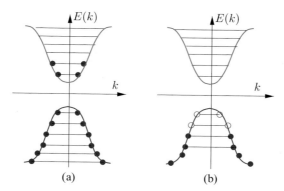

Figure 6.4. Doping of a semiconductor. (a) n type doping: the valence band is full, but a small number of electrons (full circles) occupy the conduction band. (b) p type doping: a few electrons (open circles) are missing in the valence band: one then observes a hole conduction.

does not exist in a vacuum but only in the bulk of the crystal, is called a *hole*, and holes carry a *positive charge*. It is thus quite possible to observe an electrical conductivity due to the displacement of positive charges, or "hole conductivity". This conductivity is completely beyond the reach of classical physics, as we would have to admit that the positively charged ions contribute to the conductivity, which is absurd because ions are firmly fixed to the nodes of the crystal lattice. A p-type semiconductor (p for positive) is obtained by replacing some of the silicon atoms by other atoms, for example boron atoms, which capture electrons and leave holes in the valence band. The boron atom is an *acceptor* (of electrons). Electrical conductivity is then linked to the displacement of positive charges, those of the holes. It is because of doping that semiconductors become interesting!

Box 6.3. Electrons and holes.

At the minimum of the conduction band, the energy of an electron as a function of the wave vector k has a parabolic shape: $E = \hbar^2 k^2/(2m_e^*)$. This expression defines the electron effective mass m_e^* (Box 6.1): the concavity of $E(k)$ is directed upward. On the contrary, at the maximum of the valence band, the concavity of $E(k)$ is directed downward, which would imply a negative effective mass. The passage to holes gives a positive effective mass m_h^* to the holes, with an energy $E_h(k) = \hbar^2(-k)^2/(2m_h^*)$.

The *p-n diode* (or junction) underpins the most important applications of doped semiconductors. Suppose that we place side by side two silicon

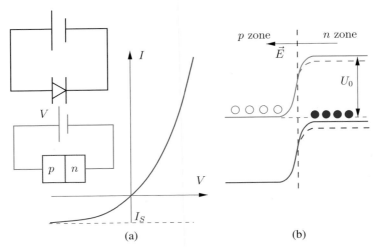

(a)   (b)

Figure 6.5. (a) Current in a *p-n* diode, or characteristic of the diode. For $V > 0$ (directly polarized diode, or forward bias) the current increases exponentially with V. For $V < 0$ (inversely polarized diode, or reverse bias), the current tends to the saturation current I_S. (b) Band diagram. In the absence of an applied voltage ($V = 0$), holes (open circles) in the neighborhood of the junction are repelled to the left and electrons (full circles) to the right by a potential barrier qV_0 whose height is close to the band gap. The height of the barrier decreases by qV (dashed lines) under application of a voltage V. $\vec{E}$ denotes the electric field.

crystals, one of them doped positively and the other one negatively, and that we apply a potential difference, or bias V. We see in Figure 6.5a that the current as a function of V is quite asymmetrical: it flows easily when $V > 0$. We then say that the diode is directly polarized (forward bias), and the current increases exponentially as a function of V. On the contrary, in the other direction, it is limited for $V < 0$ (reverse bias), to a current I_S called the saturation current. In order to explain this effect, we remark that in the absence of an external potential ($V = 0$), the junction plays the role of a potential energy barrier of height $U_0 = qV_0$, where we remind the reader that q is the *absolute value* of the electron charge. The energy barrier, whose height is close to the energy gap, repels the holes trying to flow from the *p*-zone to the *n*-zone, and conversely it prevents the electrons coming from the *n*-zone from flowing toward the *p*-zone. Indeed, the hole density is large in the *p*-zone and very low in the *n*-zone and, in the absence of a potential barrier, the densities would tend to equilibrate through a diffusion mechanism. It is the potential barrier which opposes the flow $p \rightarrow n$, by repelling the holes toward the p-zone. The equations of electrostatics allow us to compute the precise shape of this barrier, but we shall need only its qualitative shape in what follows. Let us limit ourselves to the holes, as an analogous reasoning is valid for electrons: the reason behind the choice of holes is that they have a positive charge q, so that the potential energy U has the same sign as the electric potential V. In Figure 6.5b, when a hole comes from the p-zone where the hole density is large (it is equal to that N_A of acceptors), this hole is repelled by the potential barrier and goes only rarely to the n side, as it would have to "climb" the potential barrier (Figure 6.5b). Conversely, holes are rare on the n side, but if a hole reaches the valence band, for example because of thermal motion in the vicinity of the junction, it will be immediately swept to the p-zone, as it rolls down the potential barrier. An analogous reasoning applies to electrons, which are in large number in the n-zone, but rather rare in the p-zone. When $V = 0$, there is no current, which means that the hole currents $p \rightarrow n$ and $n \rightarrow p$ equilibrate. When a positive voltage $V > 0$ is applied on the p side, the height of the barrier is decreased by qV: $U_0 \rightarrow q(V_0 - V)$, and the $p \rightarrow n$ current increases in an exponential way. When $V < 0$, the $p \rightarrow n$ current tends to be suppressed, but the $n \rightarrow p$ current is not affected and leads to the saturation current I_S. The p-n junction as described above is an example of a homojunction, because the material (for example silicon) is the same on both sides of the junction. New and spectacular effects will arise by putting two different semiconductors side by side to build a heterojunction.

Box 6.4. Hole current in a diode.

When $V = 0$, the $p \rightarrow n$ current is proportional to the probability $\exp(-qV_0/k_B T)$ for a hole to cross the barrier. The $n \rightarrow p$ current is proportional to the (low) density N_h of holes on the n side. As the currents $J_{p \rightarrow n}$ and $J_{n \rightarrow p}$ must equilibrate in the absence of an applied voltage

$$J_{p \rightarrow n} = J_{n \rightarrow p} \exp(qV_0/k_B T).$$

When $V \neq 0$, we must replace V_0 by $V_0 - V$, which influences the $p \rightarrow n$ currents but not the $n \rightarrow p$ one. The total current is then equal to

$$J_{n \rightarrow p}[\exp(qV/k_B T) - 1]$$

which vanishes, as it should, for $V = 0$. For $V > 0$, this formula is valid only if qV is smaller that the gap: $qV \lesssim E_g$.

In the wake of the pioneering work of Bardeen and Brattain, Shockley built in 1951 the first *transistor* by joining together three differently doped zones, *p*, *n* and *p*, hence the name *p-n-p* transistor, where the three elements are called the emittor, the base and the collector, respectively. The transistor allows one to generate important variations of the current between the collector and the emitter from a small current variation in the base: the transistor operates as an amplifier. The operation of a transistor, as well as that of the *p-n* diode, may be understood from an analysis of the electron and hole currents in the two junctions. The transistor which we have just summarily described is called a bipolar transistor. In most present applications, it has been replaced by the field effect transistor, mostly its MOS (Metal/Oxyde/Semiconductor) version where the three electrodes are the source, the grid and the drain. A small voltage variation applied to the grid generates important variations of the current between the source and the drain. This MOS transistor allows one to build amplifiers and switches: it is the latter property which makes the transistor an ideal tool for binary logic. The decisive technological breakthrough was the birth of integrated circuits: instead of joining together the different elements of an electronic device by wires, the transistors were directly etched by photolithography together with the passive elements (resistors, capacitors...) on a semiconductor wafer, in general a silicon plate, in order to build electronic chips. The expression "microelectronics" arose as the size of the transistors was reduced to micrometer dimensions, and even beyond. We reach today dimensions on the order of 40 nanometers, with billions of transistors etched on a single chip, about 10 millions transistors per mm^2. This downsizing of transistors is at the origin of "Moore's law", which states that the computing power of our laptops is multiplied by two every eighteen months. For different reasons, it turns out that this "law" seems to also be valid for computer memories.

6.3 Interaction with an electromagnetic field

Semiconductors, and in particular silicon, are used extensively in today's electronics, as witnessed by the expression "Silicon Valley" in California. Modern electronics is highly technological and by now rather far from fundamental physics, with the exception of some cutting edge research: one-electron transistors, carbon nanotubes, spintronics, graphene and so forth. We shall rather switch to applications of semiconductors to electro-optic devices. Semiconductors interact with electromagnetic radiation: they may absorb photons, producing an electrical current, or conversely emit photons under the action of an electrical current. The first effect is used in photon detectors and in photovoltaic cells, the second one in light emitting diodes (LEDs) and laser diodes.

When a photon with high enough energy $h\nu$ hits a semiconductor, it can bring one electron from the valence band to the conduction band, leaving a hole in the valence band: this is called the creation of an *electron–hole pair*. In order for that to occur, it is necessary that the photon energy be larger than the gap, $h\nu > E_g$, and in addition the maximum of the valence band must lie just below the minimum of the conduction band, as in Figure 6.4, which is the case for a direct gap semiconductor such as germanium. Otherwise, we deal with a so-called indirect gap semiconductor, which is the case for silicon, where the maximum of the valence band is shifted with respect to the minimum of the conduction band. Interacting with electromagnetic radiation is much easier for a direct gap semiconductor: the difficulty in the case of indirect gap semiconductors is that momentum conservation must be ensured, through phonon (or quantum of sound wave) emission, which explains why silicon is not widely used in electro-optic devices. However, even for direct gap semiconductors in bulk, interaction of photons with electron–hole pairs is weak, and not very efficient either for photon detection or for photon emission. We shall explain below how the interaction between photons and semiconductors may be boosted.

Let us concentrate on photon emission. The simplest way of building a light emitting diode is to sandwich the active zone where electron–hole pairs are formed between two slices of p-doped and n-doped semiconductors. As in Figure 6.5, the application of a positive voltage (forward bias) sweeps the holes and the electrons toward the junction, where they recombine to give photons. In other words, electrons lose an energy E_g when they jump from the conduction band to the valence band. The frequency and wavelength of the photons is determined by the gap E_g: the frequency of the emitted photon is $\nu = E_g/h$ and its wavelength $\lambda = hc/E_g$. A LED emits light with a fixed wavelength, and thus with a fixed color, for example red for a gap $E_g = 1.5\,\text{eV}$. On the other hand, a standard light bulb which, like the Sun, is a thermal source, emits white light, as its emission spectrum contains all the visible wavelengths. In order to obtain light with a fixed color, we must use color filters, and therefore lose a large part of the initial light which is absorbed by the filter. With a LED, the problem is exactly the opposite: we must combine LEDs emitting different wavelengths in order to "reconstitute" white light. However, such a light is considered as "cold" when compared to that of thermal sources. LEDs have an energetic efficiency, that is, a conversion rate of electrical energy into light energy, on the order of 20%, much higher than ordinary light bulbs, where this efficiency does not exceed 3% to 5%. The comparison is still more favorable for LEDs when we want to emit a specific color: for example, if we wish to design the tail light of a car, it is much more economical to use a LED emitting red light rather than filtering white light with a red filter.

6.4 Heterojunctions

The first generation of LEDs, whose principle we just described, were homojunctions, the semiconductor being identical on both sides of the junction. The simplicity of operation and the energetic efficiency are considerably increased if we use heterojunctions, that is, junctions where we join many different semiconductors. Indeed, the density of electron–hole pairs is much higher in a heterojunction for an identical flux of charge carriers. These heterojunctions do not exist in Nature; they are manufactured entirely artificially by epitaxy of successive monoatomic layers on a substrate, or by organo-metallic chemical vapor deposition. In order for the process to be possible, it is necessary that the cells of the semiconductors we want to join together be not too different. Figure 6.6 represents the archetype of a heterojunction, where a few dozen atomic layers of gallium arsenide (GaAs) are sandwiched between two slices of $Al_xGa_{1-x}As$: this formula tells us that a fraction x of gallium atoms in GaAs is replaced by the same fraction of aluminium atoms. Since the GaAs gap is smaller than that of the other semiconductor, this heterojunction builds a quantum well of the kind introduced in Section 4.2, whose width is on the order of 10 nanometers. Electrons are trapped in the direction of layer growth,

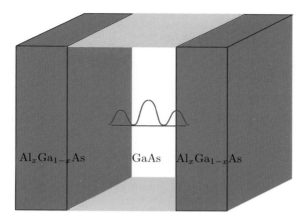

Figure 6.6.   An $Al_xGa_{1-x}As$/GaAs heterostructure, forming a quantum well. Al = aluminium, Ga = gallium, As = arsenic. Al_xGa_{1-x}: blue, GaAs: pink. The GaAs gap is smaller than that of Al_xGa_{1-x}, which allows the formation of a quantum well. The square of the wave function in the well has also been represented.

but are free to move in the perpendicular direction. We have thus built a two-dimensional electron gas. The picture described in Section 4.2 is qualitatively valid along the direction of growth. Of course, the well is not infinite, but there exist a small number of discrete energy levels. The existence of these levels leads us to distinguish between two kinds of transitions (Figure 6.7).

1. Interband transitions, which take place between an energy level in one band and an energy level in another one thanks to electron–hole recombination. The energy of a photon produced in such a transition is on the order of that of the gap (from 0.7 eV to 3 eV), but with a correction due to the levels of the well. This feature allows us to modify the photon wavelength and to adjust it on request.
2. Intraband transitions, which take place between two energy levels in the same band, and involve a single kind of charge carrier, in general an electron. These transitions are quite similar to those described in the well of Section 4.2. The typical energy differences between levels is in this case a few tenths of electron-volts, which corresponds to the infrared emission of laser diodes using quantum cascading.

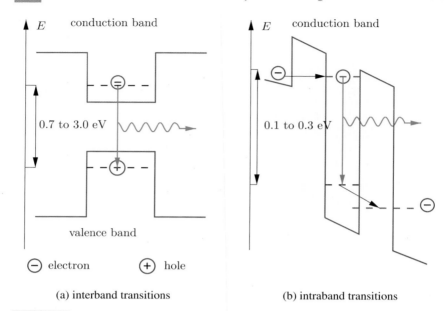

(a) interband transitions   (b) intraband transitions

Figure 6.7.   Interband and intraband transitions. (a) Interband transition: photons are emitted through a recombination of an electron–hole pair. (b) Intraband transition: an electron goes from a higher energy level of a quantum well to a lower one, while emitting a photon. The energy E is plotted on the vertical axis.

We observe that the emitted wavelengths can be adjusted practically on demand, according to an engineering based on quantum physics: we may rightly speak of *quantum engineering*. Heterojunctions are used in the fabrication of modern LEDs, whose design is displayed schematically in Figure 6.8a. The light from a LED, which corresponds to spontaneous emission, is not directional and the large value of the optical index of semiconductors implies that photons are trapped because of internal reflections, even for incident directions close to the normal. However we may take advantage of some "photon recycling": a reflected photon can itself generate an electron–hole pair.

In order to manufacture a laser diode (Figure 6.8b), and not simply a LED, we need two additional features identified in Section 4.5.

1. A mechanism of population inversion between the electrons of the conduction band and the holes of the valence band, which is obtained by increasing the applied voltage and, as a consequence, the current in the

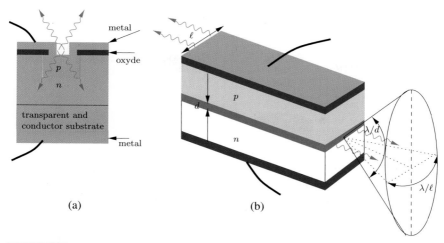

(a)   (b)

Figure 6.8. Schematic depiction of a light emitting diode and of a laser diode. (a) Light emitting diode: the substrate being transparent, light emitted backward is reflected by the downward metallic face, which increases the efficiency of the diode. (b) Laser diode: two confining layers of p- and n-type surround the active zone where the photons are produced. This active zone has width d, and is usually formed of one or many quantum wells; light is emitted in the plane of the junction. In the horizontal direction, the beam divergence is due to diffraction controlled by the witdth ℓ of the diode, typically $\ell = 5\,\mu$m, and hence an angular aperture on the order of $10°$, and in the vertical direction by the width of the active zone, typically $d = 1\,\mu$m, and hence an aperture on the order of $60°$.

junction. Below some threshold current, the junction operates as a LED, and above it as a laser diode.

2. A resonant cavity able to select a small number of modes to be amplified.

However, with a simple p-n junction, the current which is necessary for laser operation is large, and leads to overly high emission of heat. Heterojunctions solve the problem by decreasing the thickness of the active zone, which increases the electron–hole density. Moreover, the central layer with a large optical index plays the role of a very efficient wave guide for light and traps it in the amplifying zone. The operation of the diode as a laser can be generated by using the natural reflectivity of the faces obtained when cleaving the crystal: as the crystal is cleaved twice along the same crystallographic plate, the mirrors built in this way are exactly parallel, and in spite of a low reflectivity on the order of 30%, this suffices to form a resonant cavity because the low reflectivity is compensated by the high

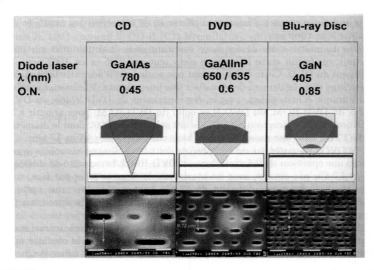

Figure 6.9. Reading CDs, DVDs and blu-ray DVDs with a laser diode. The wavelengths are given in nm, and ON (Numerical Aperture) characterizes the focusing optics. The "pits" have a depth $\lambda/4$, so that the reflected light acquires a π phase-shift. The dimensions of the pits and the distance between the tracks decreases by a factor 3 when going from a CD to a blu-ray, which multiplies the density of information by a factor of 9. Courtesy of Jean-Paul Pocholle.

density of photon emitters. As the cavity is much shorter than that of a conventional laser (Chapter 4), the intervals between the different wavelengths corresponding to the modes of the cavity are large, so that a single mode is easily selected. As of today, laser diodes may operate with threshold currents as low as a few mA; the typical dimensions of these diodes are about $500\,\mu\mathrm{m}\times 5\,\mu\mathrm{m}$. They have multiple applications: reading of bar codes, of CDs and DVDs, and, of course, optical fiber communications where one mostly uses the infrared wavelengths $1.31\,\mu\mathrm{m}$ and $1.55\,\mu\mathrm{m}$, which correspond to the minimum of absorption of optical fibers. Figure 6.9 illustrates the progress made in miniaturization: CDs are read with a laser operating in the near infrared ($\lambda = 780\,\mathrm{nm}$), and the distance between two tracks is $1.6\,\mu\mathrm{m}$, while blu-ray disks are read with a wavelength $\lambda = 405\,\mathrm{nm}$, in the near ultraviolet, with a distance between the tracks of $0.5\,\mu\mathrm{m}$. The density of information which is etched on the disk is inversely proportional to the wavelength squared: one gains a factor $(780/405)^2 \simeq 3.7$ when going from a CD to blu-ray disk, and another factor of 4 thanks to the numerical aperture ON, the ratio of the lens diameter to its focal length. The diameter of the spot on the disk is fixed by the laws of diffraction, and is given by $1.2\lambda/ON$.

6.5 Further reading

See Hey and Walters [2003], Chapters 6 and 9, and, at an advanced level, Rosencher and Vinter [2002].

7

Relativistic Quantum Physics

Up to now, we have only considered "non-relativistic" systems, where the velocities of massive particles such as electrons, protons, atoms, and so forth, are small with respect to the speed of light, c. Being massless, photons travel, of course, at the speed of light. This chapter is going to combine quantum physics with relativity, at first with special relativity (Einstein 1905) in Sections 7.1 and 7.2, and then in Section 7.3 with general relativity (Einstein 1915). In the first case, we obtain relativistic quantum field theory which is needed for elementary particle physics, and the first two sections will describe the present status of the theory which is known as the standard model of elementary particle physics, the standard model for short. The second case is that of quantum gravity and, with the example of length measurement, we shall try to understand the difficulties which this theory encounters and which have not yet been overcome today.

7.1 Relativistic quantum field theory

7.1.1 Vacuum energy

In the preceding chapters, we examined the quantum description of atoms (Chapters 4 and 5) and of electrons (Chapter 6), but we did not really address the case of photons. In this latter case, in order to obtain a quantum description, we must follow a path which is the exact inverse of that followed for electrons and atoms. In classical physics, particles are micro-billiard balls, and it was only at the beginning of the 1920s that physicists realized that they had also wave properties. As far as photons are concerned, things work in exactly the opposite way: the classical description of light, which dates back to the beginning of the XIXth century, is wave-like, and it was only in 1905 that Einstein introduced the particle aspect. In classical physics, the electromagnetic field is a numerical function of space and time, more precisely six numerical functions for the three components of the electric field and the three components of the magnetic field. In order to obtain the "particle" (photon) aspect, we must "quantize" the electromagnetic field. To understand the quantization process, it is convenient to take as an example a field confined in a cavity with reflecting walls. For simplicity, as we have already done on many occasions, we limit ourselves to a one-dimensional cavity with length L. As in the case of the vibrating string of Section 4.1, only specific oscillating modes of the field are allowed: those whose wavelengths λ obey $L = q\lambda/2$, where q is a positive integer. As for the vibrating string, the length of the cavity must be a multiple of the wavelength divided by 2. This condition on the wavelength is translated into a condition on the frequency, $v = c/\lambda = qc/2L$. One can show that a mode of the cavity behaves as a harmonic oscillator; furthermore, from the quantum study of the harmonic oscillator, one knows that the energy levels are of the form $E_n = (n + 1/2)hv, n = 0, 1, 2 \ldots$. This result has already been quoted in Section 5.3, and the energy of the $n = 0$ ground state is derived in Appendix A.4.2 from Heisenberg's inequality. We may then understand the origin of the particle aspect of the quantized field: a mode of the field with frequency v is a quantized harmonic oscillator, the energy in this mode can only take discrete values labeled by an integer n, $E_n = (n + 1/2)hv$, the energy of a single photon is hv, and the number of photons in the mode is nothing other than the integer n! The state with zero photons, $n = 0$, corresponds to an energy $E = hv/2$, the one-photon state $n = 1$ to an energy $E = 3hv/2$, the two-photon state $n = 2$ to $E = 5hv/2$, and so on.

However, the reader may at this point be completely baffled by the following observation: if the mode of frequency v contains zero photons, how is it that its energy, called its zero point energy, is $hv/2$, and not zero?

Still more baffling is the fact that there is an infinity of modes which obey $v = qc/2L$, as there is no restriction on the integer q, and the sum of all the $hv/2$ leads to an infinite energy. The state with zero photons, also called the *vacuum state* has an infinite *vacuum energy*! The reader is perfectly right if he or she thinks that all this does not make much sense, and, as of today, we do not know the final answer to this infinite energy puzzle. The generally accepted procedure, called "renormalization", consists in redefining the zero of energy by subtracting for each mode its zero point energy $hv/2$. This procedure, in which one manipulates infinite quantities, may seem mathematically questionable, but it can be made rigorous, at least in the case of special relativity. The problem remains open in general relativity, where one cannot play arbitrarily with energies, and where such a renormalization procedure does not work. The vacuum energy leads to effects which are similar to those of the cosmological constant introduced by Einstein around 1930 for the following reason. In the 1930s, astronomers believed that the Universe was static. However, this belief was contradicted by general relativity, which predicted that, starting from a static situation, the Universe would contract under the influence of gravity. In order to avoid this unwanted effect, Einstein proposed to add to his theory a cosmological constant, whose effect was to counterbalance gravity and to ensure a static Universe. Unfortunately, this remedy did no good, since the Universe turned out to be unstable. The cosmological constant was buried a few years later, when Hubble discovered that the Universe was not stationary, but expanding, and Einstein called the cosmological constant "the biggest blunder of his life". The cosmological constant made a shattering comeback in the 1990s, when astrophysicists became convinced of the acceleration of the Universe expansion, a quite surprising and unexpected discovery, given that gravity can only slow down expansion. In order to understand this point, let us use the following analogy: a rocket is launched from Earth with a velocity smaller that the escape velocity, 11 km/s. It first rises, the analogue of an expansion phase, reaches a maximum altitude and then falls back, the analogue of a contraction phase: we observe a "big bang", followed by a "big crunch". If the rocket is launched with a velocity larger than the escape velocity, it will travel to infinity, the analogue of an expansion phase lasting an infinite time, but the motion will always be slowed down by gravity;

there will never be an accelerated phase. However, the cosmological constant, which would account for the acceleration of the expansion, faces a major problem: its estimate from quantum field theory is wrong by a factor about 10^{120} if compared to its experimental value, which is by far the worst theoretical result ever obtained by physicists. There are at present many speculations but no real clue to the solution of the cosmological constant mystery.

We shall come back to gravity in the last section, and we restrict ourselves for the time being to the framework of special relativity. It is possible to be convinced of the reality of the vacuum energy thanks to the *Casimir effect*, from the name of the Dutch physicist who gave a theoretical derivation of this effect. Let us place two parallel metal plates in a vacuum at a distance L from each other (Figure 7.1). The existence of these plates modifies the configuration of the field modes with respect to the situation where they are absent, and this modification leads to a variation ΔE of the vacuum energy per unit area, which is translated into an attractive force F per unit area between the two plates

$$\Delta E = -\frac{\pi^2}{720}\frac{\hbar c}{L^3}, \quad F = \frac{\pi^2}{240}\frac{\hbar c}{L^4}. \tag{7.1}$$

The vacuum energy is infinite in the absence of plates. It is also infinite in their presence, but the difference is *finite* and measurable. Many accurate experiments have been performed in the past twenty years, which confirm the reality of the Casimir effect and agree quantitatively with (7.1) within a few percent.

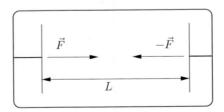

Figure 7.1. The Casimir effect. The two metal plates, placed in a vacuum at a temperature as low as possible, attract each other due to the vacuum fluctuations of the quantized electromagnetic field, or photon vacuum.

Box 7.1. Force and energy.

The force between the plates is the negative of the derivative of the energy with respect to the distance L

$$F = -\frac{\mathrm{d}\Delta E}{\mathrm{d}L}.$$

To calculate the force, we must first compute the energy in a three-dimensional space, by studying the modification brought by the plates to the boundary conditions for the electromagnetic field.

7.1.2 Virtual particles and Feynman diagrams

Relativistic quantum field theory is not limited to the electromagnetic field and to the associated particles, the photons; it deals with all kinds of particles. Before we address elementary particle physics, we recall some basic notions of special relativity. Everyone knows the famous relation $E = mc^2$, the energy–mass equivalence relation. This relation allows us to measure masses in units of energy, for example in eV/c^2 or MeV/c^2: in these units, the electron mass is $0.51\,MeV/c^2$, while that of the proton is $938\,MeV/c^2$, about $1\,GeV/c^2$. The $E = mc^2$ relation implies that massive particles can be created, provided enough energy is available. High energy photons, or γ photons, passing in the vicinity of an atomic nucleus may create electron (e^-)–positron (e^+) pairs according to the reaction

$$\gamma \rightarrow e^- + e^+.$$

The *positron* is the antiparticle associated to the electron: it has the same mass but an opposite charge. Relativistic quantum field theory predicts that an antiparticle with the same mass and an opposite charge is associated to every particle: the antiproton, for example, is the antiparticle associated to the proton, and the antineutron the antiparticle associated to the neutron. The antineutron, similar to the neutron, has a charge zero, but the sign of its magnetic moment is opposite to that of the neutron. Some particles, such as photons, are their own antiparticles. When a particle has a momentum p, its relativistic energy is given by

$$E = \sqrt{m^2c^4 + p^2c^2}. \tag{7.2}$$

We recover $E = mc^2$ for a particle at rest ($p = 0$), and $E = cp$ for a massless particle such as the photon ($m = 0$). A reaction between particles must conserve not only the total energy, but also the total momentum. This is the reason why, even for extremely high energy photons, the reaction $\gamma \rightarrow e^- + e^+$ is not possible in a vacuum: we cannot ensure both energy and momentum conservation, we need for example a nucleus for this conservation. Note also that, because of the Doppler effect, a photon which is very energetic in an inertial reference frame linked to the Earth may be much less energetic in an inertial reference frame moving with a high velocity with respect to the Earth. It is then not surprising that the reaction $\gamma \rightarrow e^- + e^+$ does not occur in a vacuum, as our description should not depend on the reference frame.

The relation (7.2) and the possibility of creating $e^- - e^+$ pairs have important consequences on the quantum behavior of an electron. Let us try to confine an electron in a small volume with linear dimensions L. Heisenberg's inequality implies that its momentum is on the order of $\hbar/L$, and from (7.2), its energy is roughly

$$E \sim \sqrt{\hbar^2 c^2 / L^2 + m_e^2 c^4},$$

where m_e is the electron mass. If we assume that L is small enough, we can neglect the electron mass and write $E \sim \hbar c/L$. Now, if $\hbar c/L \gg m_e c^2$, or equivalently $L \ll \hbar/m_e c$, we have enough energy to create $e^+ - e^-$ pairs, since we need only $2m_e c^2$ in order to do so. All this discussion implies that an electron can no longer be considered as an isolated particle if we try to confine it in a region of dimension $L \lesssim \hbar/m_e c$: this electron is then surrounded by a "cloud" of $e^+ - e^-$ pairs. The length $L = \hbar/m_e c \simeq 3.86 \times 10^{-13}$ m is called the *Compton wavelength* of the electron. At distances smaller than the Compton wavelength, the non-relativistic quantum theory of electrons breaks down, as it is unable to describe creation and annihilation of particles.

An intuitive way of understanding some results of relativistic quantum field theory is to use the temporal Heisenberg inequality, $\Delta E \Delta t \gtrsim \hbar$, where ΔE is a measure of energy fluctuations and Δt a time interval typical of the process under consideration. Although superficially reminiscent of the usual Heisenberg inequality $\Delta x \Delta p \gtrsim \hbar$, the temporal inequality has a different physical interpretation. Actually, Δx and Δp are fluctuations, and so is ΔE, but not Δt: for a physical system, x and p are dynamical variables, and so is E, but time is not a dynamical variable, it is a parameter, external to the system under consideration. As in the case of usual Heisenberg's inequalities, we use the temporal inequality in the heuristic form $\Delta E \Delta t \sim \hbar$, which implies that it is possible that energy fluctuates wildly, provided this happens during a sufficiently short time interval $\Delta t \sim \hbar/\Delta E$, so that the energy fluctuation violates the temporal inequality and cannot be observed. Physicists call such processes *virtual processes*, which are not directly observable, but whose indirect consequences are visible. The famous *Feynman diagrams* represent the most important application of virtual processes. Let us give an example, by considering the proton (p)–neutron (n) interaction, through the exchange of π-mesons, which can be charged $(\pi^{\pm})$ or neutral (π^{0}), see Figure 7.2. The reaction $p \rightarrow n + \pi^{+}$ is forbidden by relativistic energy–momentum conservation. However, we may take advantage of an energy fluctuation $\Delta E \sim \hbar/\Delta t$ during a very short time interval Δt to create a π^{+} meson, which is absorbed in the same time interval by the neutron, according to the reaction $\pi^{+} + n \rightarrow p$, so that, at the end of the day, energy–momentum *is* conserved: the global result is mere elastic proton–neutron scattering. In order to create the meson, we need an energy fluctuation $\Delta E \sim m_{\pi}c^{2}$, and then $\Delta t \sim \hbar/m_{\pi}c^{2}$. As the maximum speed of propagation is that of light, the virtual π-meson can at most travel $c\Delta t = \hbar/m_{\pi}c$, the Compton wavelength of the π-meson, so that the nuclear proton–neutron interaction must become negligible beyond that distance. The study of

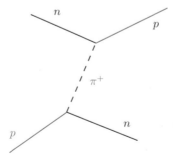

Figure 7.2. A Feynman diagram. The initial proton (red) give a neutron (blue) and a π^{+}-meson (dashed green). The π^{+}-meson is absorbed by the neutron (blue) and gives a proton (red).

nuclear forces did show that they decrease very fast beyond 1 fm (10^{-15} m), so that the Japanese physicist Yukawa was able to predict in 1935 the existence and the mass of the π-meson, around 100–200 MeV/c^2. This meson was discovered in 1947 with a mass of 140 MeV/c^2, confirming the theoretical prediction ... and awarding Yukawa a Nobel prize. One can see from this example how virtual processes may govern observable physical effects.

7.1.3   What is an elementary particle?

We learn in high school that matter is composed of molecules, molecules of atoms, atoms of electrons and nuclei, nuclei of protons and neutrons. Is there an end to this decomposition of matter into more and more elementary constituents, or does it go on to smaller and smaller scales? The first question to be asked is: how do we convince ourselves that some entity is composed of more elementary ones? If you want to decompose into elementary entities your grandfather's fob watch which you just found in an abandoned drawer, an efficient technique — which should probably be avoided — is to smash the watch with a hammer: you will discover the springs and the cogwheels which were part of the watch. Physicist do exactly the same: they break particles (the watch) by smashing them with energetic particles from an accelerator (the hammer). Let us try to break the simplest atomic nucleus, the deuteron (D), formed of a proton and a neutron. The corresponding atom, the deuterium, is an isotope of hydrogen, which is found in heavy water D_2O. Let us smash the deuteron with electrons (e). If the electron energy is low, less than 2 MeV, collisions are elastic, much as the collisions of two billiard balls: $e + D \rightarrow e + D$. The deuteron remains in one piece because the smash with the hammer was not energetic enough. For electron energies higher than 2 MeV, it may happen that the deuteron breaks into a proton and a neutron: $e + D \rightarrow e + p + n$. This is what we wanted: we have decomposed the deuteron into more elementary constituents. Let us continue the process by trying now to break the proton. For energies lower than 150 MeV, nothing happens, collisions are elastic: $e + p \rightarrow e + p$, but if the energy is increased, one observes the *creation* of new particles, charged ($\pi^{\pm}$) or neutral (π^0) π-mesons, which contrary to the π-mesons encountered in the preceding subsection are perfectly *real* particles and can be observed in detectors

$$e + p \rightarrow e + p + \pi^0,$$

$$e + p \rightarrow e + n + \pi^+.$$

By increasing further the energy, we produce more and more π-mesons or even heavier particles such as K-mesons and hyperons. In the preceding reactions, the π-mesons were not present in the proton before the collision; they were *created* at the moment when the collision took place. The proton is *not* composed of a proton and a π^0-meson, or of a neutron and a π^+-meson. Another example is that of the β-radioactivity of the neutron. As a general rule, a particle A may decay into particles $B, C, \ldots$ if its mass is larger than the sum of the masses of the final particles. The decay $A \rightarrow B + C$, for example, is possible if the masses obey $m_A \geq m_B + m_C$. The decay is then compatible with relativistic energy–momentum conservation. The neutron mass m_n is larger that the sum of the masses of the proton m_p, the electron m_e and the antineutrino $m_{\bar{\nu}}$, $m_n > m_p + m_e + m_{\bar{\nu}}$

$$m_n = 939.5 \, \text{Mev}/c^2, \quad m_p = 938.3 \, \text{Mev}/c^2, \quad m_e = 0.5 \, \text{Mev}/c^2, \quad m_{\bar{\nu}} \simeq 0$$

and the decay

$$n \rightarrow p + e + \bar{\nu}$$

is allowed. It is the simplest case of β-radioactivity: the neutron is unstable with a lifetime of about 15 minutes. However, we should not conclude that the electron preexisted in the neutron before the decay. Indeed, an electron confined in a neutron with radius of 0.8 fm would have, because of Heisenberg's inequality, a huge energy on the order of a few hundreds MeV. The neutron is not composed of a proton, an electron, and an antineutrino. It may be worth explaining why the neutron does not decay inside a stable atomic nucleus, for example, a deuteron. Why is the decay $D \rightarrow 2p + e + \bar{\nu}$ forbidden? We must look at the masses: since $m_D < 2m_p + m_e + m_{\bar{\nu}}$, the decay is forbidden by energy–momentum conservation. For the same reason, the neutron does not decay inside stable atomic nuclei.

The landscape which is suggested by relativistic quantum field theory is one where particles are created and annihilated at will, provided enough energy is available. One important consequence of this theory is the indistinguishability of quantum particles. For example, particles such as photons or electrons, are excitations of a quantized field, the electromagnetic field in the former case and the electron–positron field in the latter. As a result, all particles of the same type have a unique origin, the excitations of the corresponding quantized field, and these particles are then strictly identical (see, however, Box 7.2). Moreover, relativistic quantum field theory shows that the statistics of a particle is linked to its behavior under a 360° rotation, that is, a rotation equivalent in principle to no rotation at all. Under such a rotation, the quantum state of a boson is multiplied by $+1$, while that of a fermion is multiplied by -1. This is the famous *spin-statistics theorem*, which is easy to state, but whose proof is unfortunately extremely involved and cannot be explained intuitively.

The vacuum is a medium with intense creation and annihilation of virtual particles. Over short enough intervals, energy fluctuations create pairs of virtual particles, for example, $e^{+} - e^{-}$ pairs. Even though these particles cannot be directly observed, they have indirect and measurable effects. The creation of $e^{+} - e^{-}$ pairs, for example, shifts one of the levels of the hydrogen atom by -1.12×10^{-7} eV, a shift which has been observed experimentally with great precision. The first relativistic quantum field theory, that of the relativistic interactions between photons and electrons, called Quantum Electrodynamics (QED), was elaborated just after World War II. This theory has been experimentally verified with extraordinary precision. In addition to the shift of hydrogen energy levels, mentioned above in a particular case, this theory allows us to compute a quantity called the anomalous magnetic moment of the electron, denoted a_e, whereas $a_e = 1$ in non-relativistic quantum theory. The theoretical and experimental values are compared below:

- $a_e^{\text{exp}} = 1.001\ 159\ 652\ 180\ 8 \pm 8$ Experiment
- $a_e^{\text{th}} = 1.001\ 159\ 652\ 175 \pm 9$ Theory

The red digits give the uncertainty of the results. The theoretical uncertainty comes mainly from our imperfect knowledge of the electron charge q_e. In fact, this very precise value of a_e can be used to obtain the most accurate value of the electron charge known today. More precisely, the combination of physical constants which is involved in the calculation is the fine structure constant $\alpha = q_e^2/(4\pi\varepsilon_0 hc) \simeq 1/137$ (Box 4.3).

7.2 The standard model of particle physics

For about forty years, physicists have been able to rely on a theory of elementary particles called the *standard model* (of elementary particles). This model results from the contribution of several physicists, from roughly 1960 to 1973. Since 1973, this model has been confirmed with a precision on the order of 0.1%, mainly from experiments performed at the Large Electron Positron Collider (LEP) at CERN near Geneva, which was commissioned in 1990 and shut down in 2000 (Figure 7.3) to make room for the LHC (Figure 7.4). The standard model answers the question "What is an elementary particle?" as follows. An elementary particle is such that *it can be considered as point-like in its interactions with other particles*. In other words, the theory relies on the property that interactions take place

Figure 7.3. Aerial view of the CERN site. The red circle marks the location of the tunnel with 27 km of circumference where the LEP (Large Electron Positron), and later the LHC (Large Hadron Collider) were built, at about 100 m underground. Reproduced with CERN's authorization.

Figure 7.4. The LHC and its detectors.

at a single space-time point, and an elementary particle is structureless and indivisible. With this definition, electrons and photons are elementary, but protons and neutrons are not, as they are formed of point-like particles called *quarks*. To build protons and neutrons, we need two kinds of quarks, the up quark (u) with charge 2/3 in units of the proton charge, and the down quark (d) with charge $-1/3$. A proton is composed of two up quarks and one down quark ($2/3 + 2/3 - 1/3 = 1$), and a neutron of one up quark and two down quarks ($2/3 - 1/3 - 1/3 = 0$). Of course, antiquarks with opposite charges are associated with these two quarks, the $\bar{u}$ quark of charge $-2/3$ and the $\bar{d}$ quark with charge $+1/3$. A π^+ meson corresponds to a combination $u\bar{d}$ and a π^0 meson to a mixture of $u\bar{u}$ and $d\bar{d}$. However, at least within the present state of our knowledge, quarks do not exist as free particles and, as a consequence, we cannot break a proton into three quarks. The theory of interactions between quarks, or Quantum Chromodynamics (QCD), predicts that quarks are always "confined"; they do not exist as free particles and cannot propagate in a vacuum. Although we are not able to see them, the case for the "existence" of quarks is nevertheless unassailable, because the theory explains a large number of experimental results, has never been found to be faulty, and because there is no plausible competing theory.

Box 7.2. Relation between fields and particles.

Quantum chromodynamics shows that the relation between quantum fields (in the present case the quark-antiquark fields) and observed particles (protons and neutrons), which are called asymptotic states, is actually quite involved: the correspondence field–particle must not be viewed in a too simplistic way.

In order to continue with the standard model description, we recall that there exist four kinds of fundamental interactions, or forces.

1. The electromagnetic interactions, which govern all properties of atomic, molecular and solid state physics (Chapters 4 to 6).
2. The strong interactions, which bind together protons and neutrons in atomic nuclei.
3. The weak interactions, which are responsible for β-radioactivity, for example, neutron decay.
4. The gravitational interactions, which govern gravity on the Earth's surface and the motion of planets orbiting the Sun, are familiar in everyday life.

Gravitational interactions play strictly no role in the standard model, due to their intrinsic weakness: as the two force laws have the same spatial $1/r^2$ behavior, we can compute the ratio of the gravitational force F_{grav} to the electrostatic force F_{el} between an electron and a proton in an hydrogen atom. We find $F_{grav}/F_{el} \sim 10^{-39}$! Therefore, gravitational interactions are completely negligible in elementary particle physics; we shall come back to them in the next section. Furthermore, the standard model unifies the electromagnetic and weak interactions into electroweak interactions. The preceding classification of interactions allows us to classify the standard model particles into three categories: the leptons, the quarks, and the bosons. The *leptons* are sensitive to electroweak interactions only, while the quarks are sensitive not only to electroweak interactions, but also to strong interactions. Leptons and quarks are fermions — they are spin 1/2 particles. Spin, or intrinsic angular momentum, is a property similar to polarization, and spin 1/2 particles have two independent polarization states, for example left-handed and right-handed circular polarizations (see (11.16) in the case of photons). On the contrary, the particles which mediate the interactions, photons for electromagnetic interactions, the $W^{\pm}$ and Z^{0}-bosons for electroweak interactions, and gluons for strong interactions, are bosons — they are in fact spin 1 particles. Massless spin 1 particles such as photons have two independent polarization states, but massive spin 1 particles have three independent polarization states. Gluons, similarly as quarks, are confined and do not exist as free particles. There are three families of particles in the standard model, and the contents of the first family is summarized below.

1. Leptons: electron (e^{-}) and electronic neutrino (ν_{e}).
2. Quarks: up quark (u) and down quark (d). There are in all 6 quarks, as quarks possess a quantum characteristic, color, which is the analogue of the electric charge for strong interactions: a quark can take any of three colors. The color quantum number explains why the quantum field theory of quarks and gluons is called quantum chromodynamics (QCD). The terminology is, of course, completely conventional, and quark color has nothing to do with ordinary color!
3. The photon (γ) and the $W^{\pm}$ and Z^{0}-bosons, which are the mediators of electroweak interactions, and the gluons, which are the mediators of strong interactions between quarks. These bosons are the *gauge bosons* of the standard model, and the name finds its origin in the fact that the standard model is mathematically a gauge theory.

We must of course add all the corresponding antiparticles: the positron (e^{+}), the electronic antineutrino $(\overline{\nu}_{e})$, the up $(\overline{u})$ and down $(\overline{d})$ antiquarks. This first family is essential to everyday life. We are made of electrons and of quarks bound in atomic nuclei thanks to gluons, we see the world and receive the Sun energy thanks to photons, while neutrinos, $W^{\pm}$ and Z^{0}-bosons are necessary for the nuclear reactions which produce solar energy.

It is very surprising that the list of elementary particles does not stop with the first family. For reasons which are at present not understood, Nature decided to add two other families, which are replicas of the first one as far as leptons and quarks are concerned, while bosons are common to the three families. The second family is composed as follows.

1. Leptons: muon (μ^{-}) and muonic neutrino (ν_{μ}).

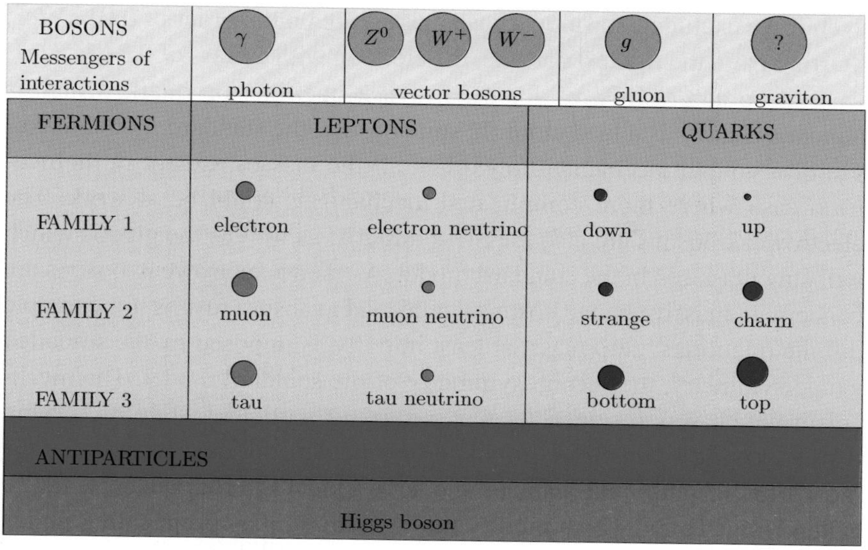

BOSONS Messengers of interactions	γ photon	Z^0 W^+ W^- vector bosons	g gluon	? graviton
FERMIONS	LEPTONS		QUARKS	
FAMILY 1	electron	electron neutrino	down	up
FAMILY 2	muon	muon neutrino	strange	charm
FAMILY 3	tau	tau neutrino	bottom	top
ANTIPARTICLES				
Higgs boson				

Figure 7.5.   The particles of the standard model.

2. Quarks: strange quark (s for "strange", charge $-1/3$) and charmed quark (c, charge $2/3$).

and their antiparticles: μ^+, $\overline{\nu}_\mu$, $\overline{s}$ and $\overline{c}$. The third family is composed of the following quarks and leptons.

1. Leptons: tau (τ^-) and τ-neutrino (ν_τ)
2. Quarks: bottom quark (b, charge $-1/3$) and top quark (t, charge $2/3$)

and their antiparticles: τ^+, $\overline{\nu}_\tau$, $\overline{b}$ and $\overline{t}$. This is summarized in Figure 7.5. There remains a last and somewhat mysterious particle, the Higgs boson, to which we shall soon return.

The standard model is grounded on two powerful symmetries. One of these symmetries implies that, initially, all spin $1/2$ particles must be massless, which is, of course, in contradiction with experiment, even for the neutrino which has a tiny but non-zero mass. The reason is that only fermions with left-handed circular polarization may interact with the charged electroweak gauge bosons, while a mass term would mix left-handed and right-handed fermions. We immediately understand that explaining quark and lepton masses will not be straightforward, as they lie in a wide interval even if we exclude neutrinos whose masses are on the order of 0.01 eV/c^2. Starting from the lightest particle, the electron with a mass of 0.5 MeV/c^2, we end up at 175 GeV/c^2 for the heaviest one, the top quark, which is almost as heavy as a lead atom. In spite of that, the standard model makes use of a unique mechanism to explain all the masses, except in the neutrino case where more complicated mechanisms could be at work. The electroweak bosons are also massless initially, as well as the gluons which will stay massless in the standard model. As far as electroweak bosons are concerned, massive gauge bosons would lead to absurd results, for example infinite quantities: zero mass, at least initially, is imposed by the so-called gauge symmetry, the second symmetry of the standard model. The mechanism which was invented to give a mass to particles, while preserving the underlying symmetries, although they become hidden, is probably the most artificial part of the standard model: we must add four quantum fields, called Higgs fields, which modify the vacuum. It is by propagating in this "Higgs vacuum" that particles, except the photon and the gluons, acquire a mass. The Higgs vacuum is endowed with a "weak charge", so that the $W^{\pm}$ and Z^0 bosons, which mediate the weak interactions, interact with this vacuum and become massive, while this vacuum is electrically neutral so that the photon remains massless. As three of the Higgs fields are absorbed by the electroweak gauge bosons, there remains one boson which appears in the form of a particle, a neutral spin zero boson, the famous *Higgs boson*. Note that the number of degrees of freedom is conserved, as it should: three Higgs degrees of freedom are eaten up by the $W^{\pm}$ and Z^0-bosons which, being now massive, possess three polarization degrees of freedom instead of two, and there remains one degree of freedom for the spin zero Higgs boson. This boson was actively searched for at LEP, and then at the Tevatron, a collider with center-of-mass energy of 2000 GeV or 2 TeV located near Chicago, but without success. As everybody knows, its discovery was finally announced at CERN in July 2012 (Figure 7.6).

Before its recent discovery as a real particle, the mass of the Higgs boson could be estimated from virtual processes. This strategy was already implemented in the case of the top quark. The top quark whose mass is 175 Gev/c^2, almost 200 times the proton mass, was difficult to produce as a real particle precisely because of this huge mass. However, the existence of the top quark influences processes in which it plays a role as a virtual

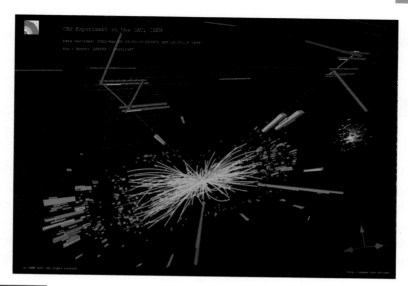

Figure 7.6. A candidate Higgs event in the CMS detector. The Higgs boson decays into two Z^0-bosons. One of them decays into two muons, detected in the muon spectrometer (green tracks). The second one decays into two electrons, whose energy is deposited in the electromagnetic calorimeter (red tracks). Reproduced with CERN's authorization.

particle, and, as early as in 1990, the LEP could predict a mass between 160 and 200 GeV/c^2, before this mass was definitively measured when the Tevatron produced the top quark as a real particle in 1995. This example illustrates the power of quantum field theory! From an analysis of processes involving the Higgs boson as a virtual particle and also from the lower limit set by LEP, the Higgs mass was predicted before July 2012 to lie between 110 and 200 GeV/c^2. The experimentally measured mass turned out to be 125 GeV/c^2. It remains to be checked that the particle found at CERN is really the standard model Higgs boson, by showing that its spin is really zero and by checking that its decay modes are those predicted by the standard model. Even though the last missing piece of the standard model has very likely been found, many physicists think that this model is not satisfactory and cannot be the last word. The standard model depends on some 20 arbitrary parameters, whose origin is not explained. The standard model could be included in a wider scheme, for example supersymmetry, which is a possible road toward a continuation of the model or, more fundamentally, string theory. However, string theory, although still promising today, encounters serious difficulties which seem far from being solved.

7.3 Quantum gravity

7.3.1 The equivalence principle

In 1915, general relativity was the outcome of intense efforts by Einstein, working practically alone to establish a relativistic theory of gravity. It is, of course, a *classical* theory, namely the theory of a classical field, the gravitational field. General relativity is mathematically complex and is beyond the scope of this book. We shall limit ourselves to a principle on which this theory is grounded, the *equivalence principle*. This principle states that in a small enough region of space-time, we cannot distinguish between an inertial force due to an acceleration and a gravitational force. An example of an inertial force is the force which tends to throw you forward when the car where you are sitting brakes suddenly; another example is the centrifugal force. Let us illustrate the equivalence principle with the following example. Consider an astronaut in a rocket located in intergalactic space. At time $t = 0$, the boosters of the rocket are switched on and the rocket is accelerated with an acceleration equal to that of gravity g ($g = 9.81$ m/s^2, Figure 7.7a). The astronaut is flung towards the back of the rocket since, with respect to the rocket (physicists say: with respect to the rocket reference frame), he is subject to an inertial force $F = -mg$, where m is the astronaut's mass and the minus sign means that the force is oriented backwards. Let us now consider the same astronaut in a motionless rocket, which is standing on the ground (Figure 7.7b), and assume that the floor gives way under his feet. The astronaut will fall towards the back of the rocket because he is subject to the force of gravity $F = -mg$. In both cases, the astronaut experiences the same feeling and, if he is not allowed to look outside, he will not be able to tell between the two situations: this is the equivalence principle.

Still inside the same rocket, let us place a source S located at an elevation $z = 0$ emitting upward single photons of frequency ν (Figure 7.7c and d). These photons are registered by a detector D located at an elevation $z = H$,

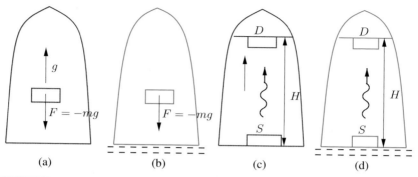

Figure 7.7. (a) A rocket is accelerated upward with an acceleration g: the astronaut is subject to a force $F = -mg$ oriented downward. (b) The rocket stands motionless on the ground: the astronaut is also subject to a force $F = -mg$ oriented downward. (c) Emission and reception of photons in the accelerated rocket. The frequency ν' received by D is lower than the emission frequency ν. (d) Emission and absorption of photons in a rocket standing motionless on the ground. The frequency received by D is again lower than the the emission frequency ν.

which is able to measure their frequency. A photon takes a time $t = H/c$ to reach the detector. In the case of the accelerated rocket, the rocket velocity, which we assume to be zero at $t = 0$, takes at time t the value $v = gt = gH/c$. The photon sees a detector which moves away, it has to "catch up" to the detector and, because of the Doppler effect (Section 5.2), the frequency ν' measured by the detector will be

$$\nu' = \nu\left(1 - \frac{v}{c}\right) = \nu\left(1 - \frac{gH}{c^2}\right). \tag{7.3}$$

We have assumed the time interval t to be short enough, so that the velocity v remains small with respect to c, and special relativity effects such as length contraction are negligible: the condition is that $gt = gH/c \ll c$, that is, $gH/c^2 \ll 1$. The equivalence principle tells us that the same frequency shift must be observed in the rocket standing on the ground, and the frequency ν' of the absorbed photon is lower than that ν of the emitted photon. This means that clocks located at a low elevation run slow with respect to those located at a higher elevation (Figure 7.8). The effect was first checked experimentally by Pound and Rebka in the 1960s, over an elevation difference of 20 m. It is essential today to take this effect into account for GPS operation: the synchronization of signals emitted from

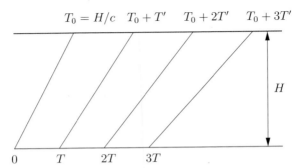

$$T_0 = H/c \quad T_0 + T' \quad T_0 + 2T' \quad T_0 + 3T'$$

Figure 7.8. The clock at elevation $z = 0$ (red) emits signals at times $0, T, 2T \ldots$. The clock at elevation $z = H$ (blue) receives them at times $T_0 = H/c, T_0 + T', T_0 + 2T', \ldots$, with $T' = T(1 + gH/c^2)$. For an observer located at elevation $z = H$, the clock at $z = 0$ runs slow.

the Earth with those emitted by satellites must be corrected in order to take into account the slowing down of earthbound clocks with respect to satellite clocks. In addition to gravitational effects, we must also take into account the time dilation of special relativity because the satellites move rapidly with respect to the Earth. The corrections due to time dilation and to gravitation are of the same order of magnitude. In the absence of these corrections, the synchronization would drift by $40\,\mu s$ a day, which may seem negligible, but in $40\,\mu s$ light travels $12\,km$! Another consequence of the equivalence principle is that a light beam propagating horizontally is curved downward by the Earth's gravitational field: light "falls"! It is easy to understand this effect by examining the trajectory of a light beam in an accelerated rocket.

7.3.2 The Planck scale

Another way of showing the gravitational slowing down of clocks is based on energy considerations. When it leaves the source S, the photon has an energy $h\nu$. If we take our inspiration from $E = mc^2$, we may attribute to the photon a "mass" $m_{\mathrm{phot}} = h\nu/c^2$. When reaching the elevation H, the photon has lost a gravitational energy $m_{\mathrm{phot}}gH = h\nu gH/c^2$, and its energy at elevation H is $h\nu(1 - gH/c^2) = h\nu'$, so that we recover for ν' a result which agrees with that deduced from the Doppler effect (7.3). The quantity

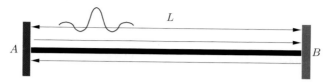

Figure 7.9. Measurement of the length L of the ruler AB. A photon represented by a wave packet (green) is sent from A. It is reflected by a mirror in B and we measure the time T of the round trip. By definition, the length is $L = cT/2$.

$P = gH$ is the gravitational potential, and we may sum up the preceding argument as follows: if the gravitational potentials of two clocks differ by P, then the clock which is located in a lower gravitational potential runs slow by a factor $(1 + P/c^2)$. We now use this result in order to derive an ultimate limit to the accuracy of length measurements. Let us assume that we wish to determine the length of a ruler by measuring the time necessary for a round trip of a photon: a photon is emitted at time $t = 0$ from one of the extremities, A, of the ruler, it is reflected at extremity B by a mirror and detected at time $t = T$ at point A after a round trip. We can use as a *definition* $L = cT/2$. In addition to experimental errors which we shall neglect, there exist two *fundamental* sources of uncertainty on the measurement result.

1. If λ is the photon wavelength, the length of the corresponding wave packet cannot be shorter than λ (Figure 7.9), and the precision of the measurement is at best λ.
2. The gravitational potential P created by the photon along the ruler is $\sim m_{\text{phot}} G/L$, where G is the gravitation constant

$$P \sim \frac{h\nu}{c^2} \frac{G}{L} = \frac{G\hbar}{Lc\lambda}.$$

This potential is a second source of uncertainty, as it influences the operation of clocks used to measure the round trip time T.

To sum up, the fundamental uncertainty ΔL on the measurement of L, which cannot be reduced whatever the accuracy of device used in the experiment, is

$$\Delta L \sim \lambda + \frac{PL}{c^2} = \lambda + \frac{G\hbar}{\lambda c^3} = \lambda + \frac{L_{\text{P}}^2}{\lambda}, \tag{7.4}$$

where $L_P = \sqrt{G\hbar/c^3}$ is *Planck's length*. In order to determine the minimum uncertainty, we must examine the curve giving $\lambda + L_P^2/\lambda$ as a function of λ. We see that the curve tends to infinity for $\lambda \to 0$, and also for $\lambda \to \infty$. There exists a minimum at $\lambda \simeq L_P$, and the minimum uncertainty is then $\Delta L \simeq L_P$. We cannot measure lengths with an accuracy better than $L_P = 1.6 \times 10^{-35}$ m. In other words, distances smaller than L_P have no meaning if we try to combine quantum physics and general relativity. Proponents of loop quantum gravity conclude from this statement that space-time has a granular structure for lengths smaller than Planck's length.

Box 7.3. Gravitational constant and gravitational potential.

The gravitational constant G is defined by the attractive force F between two point-like masses m and m' at a distance r from each other : $F = Gmm'/r^2$. The gravitational potential energy of a mass m is $U = -Gmm'/r = Pm$, where P is the gravitational potential created by mass m', $P = -Gm'/r$. Since L is the ruler length, the gravitational potential created by the photon is $P \sim Gm_{\text{phot}}/L$.

The *Planck mass* $M_P = \sqrt{\hbar c/G}$ is the mass of a particle whose Compton wavelength is L_P, and the Planck energy is $E_P = M_P c^2 = \sqrt{\hbar c^5/G}$. Its numerical value is 1.2×10^{19} GeV, an energy 10^{15} times that achievable by the LHC. At distances smaller than L_P, or at energies larger than E_P, we can no longer neglect quantum effects in general relativity. However, applying the techniques of quantum field theory leads to infinities which we cannot at present interpret in a meaningful way: general relativity is "non-renormalizable". Two main avenues are being explored today: string theory and loop quantum gravity. In both cases, we must seriously modify our previous conceptions, and the modifications have not been fully mastered. In the case of strings, there exists a minimum dimension for the particles, which are tiny strings with dimensions on the order of L_P, and in the case of loop quantum gravity, space-time takes a granular structure. However, in spite of impressive theoretical progress, we are still far from a quantum theory of gravity, if such a theory exists at all!

7.4 Further reading

There are many popular books on elementary particle physics and string theory, among them Green [1999], Randall [2011] or Gubser [2010]. For a critical view of string theory, see Smolin [2006]. At an intermediate level, see Quigg [2006] for elementary particle physics, Greene [2006] for string theory and Adler [2006] for gravity. Experimental results for the Casimir effect are reviewed in Mohiden and Roy [1998] and neutrinos in Wark [2005]. A wonderful introduction to special and general relativity is given by Taylor and Wheeler [1963].

8

Towards a Quantum Computer?

In everyday life, practically all the information which is processed, exchanged or stored is coded in the form of discrete entities called *bits*, which take two values only, by convention 0 and 1. With the present technology for computers and optical fibers, bits are carried by electrical currents and electromagnetic waves corresponding to macroscopic fluxes of electrons and photons, and they are stored in memories of various kinds, for example, magnetic memories. Although quantum physics is the basic physics which underlies the operation of a transistor (Chapter 6) or of a laser (Chapter 4), each exchanged or processed bit corresponds to a large number of elementary quantum systems, and its behavior can be described classically due to the strong interaction with the environment (Chapter 9). For about thirty years, physicists have learned to manipulate with great accuracy individual quantum systems: photons, electrons, neutrons, atoms, and so forth, which opens the way to using two-state quantum systems, such as the polarization states of a photon (Chapter 2) or the two energy levels of an atom or an ion (Chapter 4) in order to process, exchange or store information. In § 2.3.2, we used the two polarization states of a photon, vertical (V) and horizontal (H), to represent the values 0 and 1 of a bit and to exchange information. In what follows, it will be convenient to use Dirac's notation (see Appendix A.2.2 for more details), where a vertical polarization state is denoted by $|V\rangle$ or $|0\rangle$ and a horizontal one by $|H\rangle$ or $|1\rangle$, while a state with arbitrary polarization will be denoted by $|\psi\rangle$. The polarization states of a photon give one possible realization of a *quantum bit*, or for short a *qubit*. Thanks to the properties of quantum physics, quantum computers using qubits, if they ever exist, would outperform classical computers for some specific, but very important, problems.

In Sections 8.1 and 8.2, we describe some typical quantum algorithms and, in order to do so, we shall not be able to avoid some technical developments. However, these two sections may be skipped in a first reading, as they are not necessary for understanding the more general considerations of Sections 8.3 and 8.4.

8.1 | Bits and quantum logic gates

8.1.1   Classical bits and quantum bits

Before addressing the qubit case, let us briefly recall how we use classical bits to write integers. As a simple example, let us consider the integers between 0 and 7, for which we use the boldface notation $\mathbf{x}$. Three bits suffice to write integers from 0 to 7, if we use the binary notation

$$\mathbf{0} = \{000\}, \quad \mathbf{1} = \{001\}, \quad \mathbf{2} = \{010\}, \quad \mathbf{3} = \{011\}$$

$$\mathbf{4} = \{100\}, \quad \mathbf{5} = \{101\}, \quad \mathbf{6} = \{110\}, \quad \mathbf{7} = \{111\}.$$

Observe the difference in notation for an integer $\mathbf{x}$ and for the value 0 or 1 of a bit: $x = 0$ or $x = 1$. The numerical value of $\mathbf{x} = \{x_2 x_1 x_0\}$ is

$$\mathbf{x} = 4x_2 + 2x_1 + x_0 = 2^2 x_2 + 2x_1 + x_0.$$

The integer 6, for example, is decomposed as

$$\mathbf{6} = 4 + 2 + 0 \text{ so that } x_2 = 1, \quad x_1 = 1, \quad x_0 = 0.$$

Generally speaking, an integer $\mathbf{x}$ between 0 and $2^n - 1$, where n is an integer, writes

$$\mathbf{x} = 2^{n-1} x_{n-1} + 2^{n-2} x_{n-2} + \cdots + 2x_1 + x_0 = \{x_{n-1} x_{n-2} \cdots x_1 x_0\}, \tag{8.1}$$

where $x_i = 0$ or $x_i = 1$. The translation into qubits is easy. Assume that we have at our disposal three photons to write as above an integer between 0 and 7. In order to write **6**, for example, we take the first photon in the state of vertical polarization $|0\rangle$, the second one in the state of horizontal polarization $|1\rangle$ and the third also in the state of horizontal polarization $|1\rangle$. We shall write this three photon state as $|1\rangle|1\rangle|0\rangle$ (note that the first qubit is located on the right), and the 8 integers between 0 and 7 will be represented by the polarization states of our three photons

0 : $|0\rangle|0\rangle|0\rangle$,   **1** : $|0\rangle|0\rangle|1\rangle$,   **2** : $|0\rangle|1\rangle|0\rangle$,   **3** : $|0\rangle|1\rangle|1\rangle$

4 : $|1\rangle|0\rangle|0\rangle$,   **5** : $|1\rangle|0\rangle|1\rangle$,   **6** : $|1\rangle|1\rangle|0\rangle$,   **7** : $|1\rangle|1\rangle|1\rangle$.

The correspondence with classical bits is obvious. A state such as $|1\rangle|1\rangle|0\rangle$ is called a *tensor product* of the three individual states $|1\rangle$, $|1\rangle$ and $|0\rangle$ (Appendix A.3.2). Let us now move on to the general case, assuming that we wish to store in a quantum register an integer **x** between 0 and $2^n - 1$. In order to do so, we shall need n qubits, and the integer **x** will be represented by

$$|\mathbf{x}\rangle = |x_{n-1}\rangle|x_{n-2}\rangle \cdots |x_1\rangle|x_0\rangle. \tag{8.2}$$

For the time being, we have not seen much of a difference between writing integers with either classical bits or with qubits. Differences will appear right now, if we remember the properties of photon polarization. Let us come back to Chapter 2, Figure 2.1a, where the (linear) polarization of the photon incident on the polaroid makes an angle θ with the vertical axis. The corresponding polarization state is a linear superposition of a vertical polarization state $|0\rangle$ and a horizontal one $|1\rangle$ with weights $\cos \theta$ and $\sin \theta$

$$|\psi\rangle = \cos \theta |0\rangle + \sin \theta |1\rangle. \tag{8.3}$$

We recall the physical interpretation of the weights $\cos \theta$ and $\sin \theta$: if we let a photon of polarization $|\psi\rangle$ go through a polaroid with a vertical (resp. horizontal) axis, the photon will be transmitted with a probability $\cos^2 \theta$ (resp. $\sin^2 \theta$), which is nothing other than Malus's law (Appendix A.2.2) for photons. One particular case of (8.3), widely used in Chapter 2, is that of diagonal $|D\rangle$ and anti-diagonal states $|A\rangle$ polarized at $\theta = \pm 45°$

$$|D\rangle = \frac{1}{\sqrt{2}}(|0\rangle + |1\rangle), \quad |A\rangle = \frac{1}{\sqrt{2}}(|0\rangle - |1\rangle). \qquad (8.4)$$

While a classical bit can take only two values, 0 and 1, a qubit $|\psi\rangle$ can take infinitely many values, as θ may vary continuously between 0 and π. Actually, we need *two angles*, θ and ϕ, for the full description of a qubit, because (8.3) describes only linear polarizations: a qubit is represented geometrically by a point on the surface of a sphere of unit radius, the Poincaré–Bloch sphere (Figure 8.1). A point on this sphere is parametrized by two angles, the latitude (or polar angle) θ, $0 \le \theta < \pi$, and the longitude (or azimuthal angle) ϕ, $\phi \le 0 < 2\pi$. As the angles θ and ϕ vary in a continuous way, we might be tempted to conclude that a qubit contains much more information than a classical bit, in fact an infinite amount of information! However, if we measure a qubit, we must use a definite orientation of a polaroid (or of a PBS), for example V or D. The result is by convention 0 if the photon is transmitted, 1 if it is stopped, whatever the orientation. Measuring a qubit can give two results, and two results only, and our hope to extract more information from a qubit than from a classical bit is unfounded. This

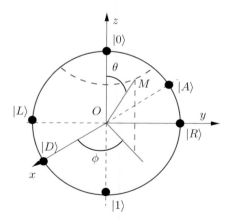

Figure 8.1. The Poincaré–Bloch sphere has radius one. A point on the sphere is represented by its latitude θ and its longitude ϕ. The polarization state $|\psi\rangle$ (Box 8.1) is in one-to-one correspondence with a point on the sphere. The points corresponding to vertical polarization $|0\rangle$ and to horizontal polarization $|1\rangle$, to the polarizations at $\pm 45°$ $|D\rangle$ and $|A\rangle$, and finally to the right-handed $|R\rangle$ and left-handed $|L\rangle$ circular polarizations (11.18), are displayed on the surface of the sphere.

pessimistic remark is confirmed by a theorem proved by Holevo: N qubits do not carry more information than N classical bits. The reasons why quantum computers may outperform classical ones are not at all obvious, and it is because of specific properties of quantum physics: superposition and entanglement, that quantum computers may turn out to be useful in some very important, but as of today limited, number of problems.

Box 8.1. Quantum state of a qubit.

The most general state vector $|\psi\rangle$ of a qubit is

$$|\psi\rangle = \cos\frac{\theta}{2}\,|0\rangle + e^{i\phi}\sin\frac{\theta}{2}\,|1\rangle,$$

which establishes a one-to-one correspondence between $|\psi\rangle$ and a point on the Poincaré–Bloch sphere parametrized by the angles θ and ϕ. We note that $|\psi\rangle$ involves the angle $\theta/2$. The case $\phi = 0$ (or $\phi = \pi$) corresponds to a linear polarization, the case $\phi \neq 0$ to an elliptic polarization. The particular cases $\phi = \pm\pi/2$ give circular polarizations.

8.1.2   Quantum logic gates

A computation done with a quantum computer is sketched in Figure 8.2, where n qubits are prepared in state $|0\rangle$ at time $t = 0$: this is the *preparation* stage of the quantum system. Then, the qubits undergo a quantum evolution which performs the calculations which are needed, for example, that of a function. The main difficulty is that the qubits must be isolated as well

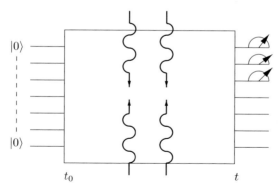

Figure 8.2.   Schematic depiction of the basic principles of quantum computing. n qubits are prepared in the state $|0\rangle$. They undergo a quantum evolution in the space spanned by the n-qubits from time $t = t_0$ to time t. The wiggly arrows represent qubit interactions with classical electromagnetic fields. A measurement of the qubits (or of a subset of the qubits, the first three in this figure) is made at time t. Time flows from left to right.

as possible from their environment, as otherwise the evolution would take place in a space larger than that spanned by the n qubits. In such a case, the reliability of the computation would deteriorate due to the decoherence phenomenon, which will be examined in the next chapter. The only interactions which are allowed are those with *classical* fields, in general electromagnetic fields, which are compatible with a quantum evolution restricted to the space spanned by the qubits. Once the evolution is completed, a measurement is made at time t on the qubits (or a subset of them), which gives the result of the calculation. Note that it is impossible to perform any observation on the qubits during the quantum evolution, that is, between times t_0 and t, while nothing prevents us from observing classical bits during an intermediate phase of a calculation run on a ordinary computer.

Quantum computing operations are implemented thanks to quantum logic gates (Figure 8.3). The simplest gates are one-qubit gates, which transform the state of one qubit into another state of the same qubit. By analogy to integers, gates will be written using boldface letters. Let us give two important examples of one-qubit gates. The **X**-gate exchanges the $|0\rangle$ and $|1\rangle$ states.

$$\mathbf{X}|0\rangle = |1\rangle, \quad \mathbf{X}|1\rangle = |0\rangle, \tag{8.5}$$

while the Hadamard gate transforms the $|0\rangle$ and $|1\rangle$ states into states $|D\rangle$ and $|A\rangle$ polarized at $\pm 45°$

$$\mathbf{H}|0\rangle = \frac{1}{\sqrt{2}} (|0\rangle + |1\rangle) = |D\rangle,$$

$$\mathbf{H}|1\rangle = \frac{1}{\sqrt{2}} (|0\rangle - |1\rangle) = |A\rangle. \tag{8.6}$$

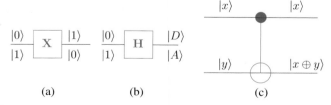

(a)   (b)   (c)

Figure 8.3. Quantum logic gates. (a) The **X** gate, which exchanges states $|0\rangle$ and $|1\rangle$. (b) The Hadamard gate **H** transforms $|0\rangle$ into $|D\rangle$ and $|1\rangle$ into $|A\rangle$. (c) Controlled-not gate **cNOT**: the control qubit is $|x\rangle$ (green line) and the target qubit is $|y\rangle$ (blue line).

The action of a Hadamard gate is that of a beamsplitter. The $\mathbf{X}$ and $\mathbf{H}$ operations squared give the unit operator: $\mathbf{X}^2 = \mathbf{H}^2 = 1$, and are then identical to their inverse. Another gate which is of major importance is the **cNOT**-gate, or controlled-not gate, which is a two-qubit gate, where the first qubit, $|x\rangle$, is the control qubit, and the second, $|y\rangle$, the target qubit. When applied to a two-qubit state, the **cNOT** gate changes nothing if the control qubit is in the $|0\rangle$ state,

$$\mathbf{cNOT}\,(|0\rangle|0\rangle) = |0\rangle|0\rangle, \quad \mathbf{cNOT}\,(|0\rangle|1\rangle) = |0\rangle|1\rangle,$$

but it exchanges $|0\rangle$ and $|1\rangle$ for the target qubit if the control qubit is in the state $|1\rangle$

$$\mathbf{cNOT}\,(|1\rangle|0\rangle) = |1\rangle|1\rangle, \quad \mathbf{cNOT}\,(|1\rangle|1\rangle) = |1\rangle|0\rangle.$$

Using the modulo 2 addition already encountered in Chapter 2

$$0 \oplus 1 = 1 \oplus 0 = 1, \quad 0 \oplus 0 = 1 \oplus 1 = 0$$

we can summarize the action of the **cNOT** gate as

$$\mathbf{cNOT}(|x\rangle|y\rangle) = |x\rangle|(x \oplus y)\rangle, \tag{8.7}$$

where x is the value of the control qubit ($x = 0$ or $x = 1$), and y that of the target qubit ($y = 0$ or $y = 1$). The controlled-not gate is the equivalent of the exclusive-OR of classical boolean logic. The importance of these gates comes from the following property: any evolution of an ensemble of qubits can be decomposed into a product of a small number of one qubit gates and of the **cNOT** gate or, more precisely, any evolution can be approached within an arbitrary accuracy by combining a limited number of these gates.

Thanks to the property $y \oplus x \oplus x = y$, the **cNOT** gate is *reversible*, and this property holds for all operations of quantum logic: given the state resulting from an evolution, we can recover the initial state by applying the inverse evolution. In other words, there is a one-to-one correspondence between the initial and final state. This is not true of operations of classical logic, which contain *irreversible* logic gates, such that two different initial states may lead to the same final state. However, we know how to transform any classical circuit into a reversible one, at the price of a limited increase of the computing time: any classical algorithm may be converted into a quantum one using essentially the same number of gates.

8.2 Quantum algorithms

8.2.1 Calculation of a function

Schematically, a quantum algorithm operates in the following way: an input register of n qubits stores an integer $\mathbf{x}$, $0 \leq \mathbf{x} \leq 2^n - 1$ and an output register of m qubits stores an integer $\mathbf{y}$, $0 \leq \mathbf{y} \leq 2^m - 1$. In order to give an elementary example, we examine the case where the input and output register both contain two qubits: each of them can store integers $\mathbf{x}$ and $\mathbf{y}$ between 0 and 3. Let us assume that a function $f(\mathbf{x})$ is given for the four values of $\mathbf{x}$ by

$$f(\mathbf{0}) = 2, \quad f(\mathbf{1}) = 3, \quad f(\mathbf{2}) = 1, \quad f(\mathbf{3}) = 0. \tag{8.8}$$

It is straigthforward to build a quantum logical circuit which computes this function. The circuit is drawn in Figure 8.4, where we recall that the input and output circuits store two qubits. This circuit has the following action on an initial state $|x_1\rangle|x_0\rangle|y_1\rangle|y_0\rangle$, where x_1 and x_0 are stored in the input register, y_1 and y_0 in the output register

$$|x_1\rangle|x_0\rangle|y_1\rangle|y_0\rangle \rightarrow |x_1\rangle|x_0\rangle|(y_1 \oplus x_1 \oplus 1)\rangle|(y_0 \oplus x_1 \oplus x_0)\rangle. \tag{8.9}$$

If the function $f(\mathbf{x})$ is defined by (8.8), then the action of the circuit is summarized by

$$|\mathbf{x}\rangle|\mathbf{y}\rangle \rightarrow U_f|\mathbf{x}\rangle|\mathbf{y}\rangle = |\mathbf{x}\rangle|[\mathbf{y} \oplus f(\mathbf{x})]\rangle, \tag{8.10}$$

where $\oplus$ is the addition modulo 2 *without carry over*. In particular, if $\mathbf{y} = 0$

$$|\mathbf{x}\rangle|\mathbf{0}\rangle \rightarrow U_f|\mathbf{x}\rangle|\mathbf{0}\rangle = |\mathbf{x}\rangle|f(\mathbf{x})\rangle,$$

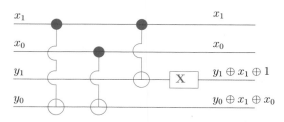

Figure 8.4. An elementary quantum circuit with three **cNOT** gates (x_1 and x_0 = control bits, green lines, y_1 and y_0 = target bits, blue lines) and a one-qubit gate $\mathbf{X}$, which computes the function $f(\mathbf{x})$ (8.8).

which gives the desired result $f(\mathbf{x})$. One can check it, for example, for $\mathbf{x} = 1$ $(x_1 = 1, x_0 = 0)$: given $x_1 \oplus 1 = 1$ and $x_1 \oplus x_0 = 1$ we deduce $f(\mathbf{1}) = \mathbf{3}$. The transformation U_f represents a particular case of the quantum evolution in Figure 8.4. For a generic function $f(\mathbf{x})$, the quantum evolution U_f is built by generalizing the circuit of Figure 8.4 to n and m qubits instead of two in the input and output registers.

8.2.2 The Deutsch algorithm

In the final scene of Howard Hawk's movie, "Only Angels Have Wings", after all sorts of incidents which have nothing to do with quantum computing, Bonnie (Jean Arthur) asks Geoff (Cary Grant) whether she should leave by boat or stay with him in the small airfield of Columbia from which he takes off to carry mail across the Andes Mountains (the action takes place in 1938). Just as he is leaving for a flight, Geoff flips a coin and tells Bonnie: "Heads you stay and tails you leave!". After Geoff has left, furious at having be played by the flip of a coin, Bonnie packs her suitcase but she has a look at the coin...and discovers that it has two heads. In order to check whether the coin was fair or not, she had to look at both faces of the coin. The Deutsch algorithm gives the result in one operation only!

Translated into mathematical terms, Bonnie's problem is the following one: she is given a function $f(x)$, $x = 0$ or $x = 1$, and $f(x)$ also takes two values 0 and 1. We need not use boldface letters as x and $f(x)$ take the values 0 and 1 only. We wish to get the following information: are the two values of $f(x)$ identical, $f(0) = f(1)$ (the coin is rigged with two heads or two tails), or $f(0) \neq f(1)$ (the coin is fair with one heads and one tails)? With a classical computer, two operations are necessary, because we need to compute $f(0)$ *and* $f(1)$ or, in other words, we need a table of values of $f(x)$. The strategy of the quantum computer in order to obtain the result $f(0) = f(1)$ (constant function) or $f(0) \neq f(1)$ (balanced function) in a single operation, will be to bypass the explicit calculation of a table of values of f and to make a direct comparison.

The Deutsch algorithm is implemented thanks to the circuit in Figure 8.5, with one qubit input and output registers. Of course, this example is too elementary to be of any practical value, but it is the simplest example which illustrates the concept of quantum parallelism. The input

Figure 8.5. The Deutsch algorithm. The qubit of the input register is initially in the state $|0\rangle$, while that of the output register is in $|1\rangle$. A Hadamard gate transforms them into states $|D\rangle$ and $|A\rangle$ (8.4) respectively. After a quantum evolution U_f, a measurement of the input register is performed in the configuration (or basis) $\{DA\}$.

register being initially in state $|0\rangle$ and the output register in state $|1\rangle$, application of a Hadamard gate to the two qubits leads to

$$(\mathbf{H}|0\rangle)(\mathbf{H}|1\rangle) = \frac{1}{2}(|0\rangle + |1\rangle)(|0\rangle - |1\rangle) = \frac{1}{2}\left(\sum_{x=0}^{1}|x\rangle\right)(|0\rangle - |1\rangle).$$

$$(8.11)$$

The quantum evolution U_f on this state changes nothing if $f(x) = 0$, but if $f(x) = 1$, it changes the sign of the state because, as implied by (8.10)

$$U_f[|x\rangle(|0\rangle - |1\rangle)] = |x\rangle(|0 \oplus 1\rangle - |1 \oplus 1\rangle) = -|x\rangle(|0\rangle - |1\rangle),$$

that is

$$(|0\rangle - |1\rangle) \rightarrow (-1)^{f(x)}(|0\rangle - |1\rangle).$$

The state of the input register is then

$$\frac{1}{\sqrt{2}}((-1)^{f(0)}|0\rangle + (-1)^{f(1)}|1\rangle) = (-1)^{f(0)}|D\rangle \quad \text{if } f(0) = f(1)$$

$$= (-1)^{f(0)}|A\rangle \quad \text{if } f(0) \neq f(1).$$

If the qubits are carried by the polarization states of a photon, the final step is to measure the input qubit with a polaroid (or a PBS) oriented at 45°. If the photon is transmitted, it means that the function is constant, $f(0) = f(1)$), and if it is stopped, it means that the function is balanced, $f(0) \neq f(1)$. This example illustrates quantum parallelism which allows us to make a comparison without explicitly calculating the values of the function.

8.2.3 Quantum parallelism

States defined by (8.3) form a basis of a 2^n-dimensional space called the computational basis, and we might be tempted to conclude from the super-position principle that a n-qubit register can store 2^n states at the same time. However, much as in the case of a single qubit, a measurement will give a unique result, and the challenge of quantum computing is to make use of superposition and entanglement in order to exploit this stored information which, in principle, grows exponentially. A state of the input register where all the qubits are in the state $|0\rangle$ will be denoted by $(|0\rangle)^n$, and similarly $(|0\rangle)^m$ for the output register. Applying a Hadamard gate on the input registers generalizes (8.11) and yields a superposition of values of $\mathbf{x}$ from 0 to $2^n - 1$

$$(\mathbf{H}|0\rangle)^n = \frac{1}{2^{n/2}} \sum_{\mathbf{x}=0}^{2^n-1} |\mathbf{x}\rangle,$$

and, given a function $f(\mathbf{x})$, the quantum evolution generalizes (8.10)

$$U_f[(\mathbf{H}|0\rangle)^n(|0\rangle)^m] = \frac{1}{2^{n/2}} \sum_{\mathbf{x}=0}^{2^n-1} |\mathbf{x}\rangle|f(\mathbf{x})\rangle.$$

In principle, this state contains the 2^n values of $f(\mathbf{x})$, which are not nec-essarily all different. For example, if $n = 100$, it contains the $\sim 10^{30}$ values of the function! It is this exponential growth of the number of states which allows quantum parallelism to deal with problems which would be intractable on a classical computer. A measurement of the output register gives, of course, only one of the values of $f(\mathbf{x})$, but it is nevertheless possible to extract useful information on *relations* between these values, the price to be paid being the loss of the individual values, as we have seen in the case of the Deutsch algorithm. *The art of quantum computing is to build an interference scheme such that the result stands out of a noisy background with a reasonable probability.* In this respect, the Deutsch algorithm is mis-leading, because it gives the desired result with certainty. More complex quantum algorithms can only give the result with some probability: they are *probabilistic algorithms*. It is then necessary to check the results with a classical computer, but that can be done quite fast. It is the search for a solution which is time consuming.

There exist at present two main classes of quantum algorithms. The first class, represented by the Grover algorithm, allows us to speed up the calculations quadratically. The goal of the algorithm is to search for an entry in an unstructured data basis, for example, to find a name in a phone directory when we only know the phone number. If the directory contains N entries, we need about $N/2$ trials before finding the right name. Grover's algorithm gives the answer in about $\sqrt{N}$ operations: if the directory contains 10 000 entries, we need 100 trials instead of 5000. This first class of algorithms exploits superposition, but not entanglement, while the other class, to which the Shor algorithm belongs, makes essential use of entanglement. This latter algorithm factors an integer N into primes. The difficulty of implementing this operation on a classical computer underpins the security of RSA encryption (Chapter 2 and Appendix A.2.1). The concrete implementation of Shor's algorithm would immediately sign the death certificate of almost all processes currently available for the security of data transmission. Today, the best algorithm running on a classical computer requires a computing time (or, equivalently, a number of operations) growing with N as $\sim \exp[1.9 \ln^{1/3} N \ln \ln^{2/3} N]$, where $\ln N$ is the logarithm of N, about the number of digits of N. As this computing time increases faster than any polynomial in $\ln N$, the number of bits $n \sim \ln N$ which characterizes the size of the problem, it has been conjectured that this factoring is a problem of exponential complexity, also called an "intractable" problem (see the next Section). On the contrary, the problem becomes of polynomial complexity only (a "tractable" problem) with Shor's algorithm, where the computing time grows only proportionally to $\sim (\ln N)^3$. It may also happen that quantum computers become useful in solving problems of condensed matter physics, such as that of strongly correlated fermions, but it is somewhat frustrating that after almost twenty years, no really novel quantum algorithm has been added to Grover's and Shor's.

As with classical computers, errors may creep into storage or processing of quantum information, and we must devise error-correcting codes. Classical error-correcting codes rely on redundancy: for example, one makes three copies of each bit and uses a majority rule to find the correct value. If the probability of an error is 1%, then the probability of the same error is only about 10^{-4} after correction. Classical error-correcting codes cannot be transposed to quantum algorithms, because the no-cloning theorem forbids us to copy an unknown qubit and the values of the qubit must not be tampered with during the computation. Also, the only possible error on a classical bit is the substitution of 0 by 1, or vice-versa, while the parameters of a qubit varies in a continuous way on the Poincaré–Bloch sphere. The uncontrolled noise which is at the origin of the errors may lead to continuous variations of the angles θ and ϕ (Box 8.1). It is quite remarkable that we can nevertheless control all possible sources of errors with a limited number of basic corrections, and efficient quantum error correcting codes have been devised, for example that of Steane, which uses a 7 qubit redundancy.

8.3 Quantum algorithms and algorithmic complexity

Quantum algorithms challenge some statements of classical algorithmics, when we ask the question of *algorithmic complexity*: what computing resources are needed to perform a calculation? A general idea is that some problems may be solved in a number of computational steps $\mathcal{N}$ which is a polynomial in the number of bits that measures the size of the problem: for example, if we wish to multiply two n-digit numbers in binary notation, the necessary number of steps is a polynomial in n, in this case $\mathcal{N} \sim n^2$. A much less trivial example is that of primality: what is the number of computational steps necessary to prove that an integer is a prime? It was shown in 2002 that this problem is of polynomial complexity only. On the contrary, experience suggests that other problems require a number of computational steps growing faster than any power of n for $n \gg 1$: for example, $\exp(n)$, $\exp(n^{1/3})$ or $n^{\ln n}$. Speaking somewhat loosely, one often terms these problems as having exponential complexity.

Turing defined a class of machines known nowadays as *Turing machines*, which allowed mathematicians to address the notion of complexity of an algorithm. He showed that there exist machines, called universal Turing machines, of which he proposed an example, that have the ability to simulate the operation of any other Turing machine. It was discovered afterwards that all computational models could be simulated by a Turing machine using a computational time that is polynomial in that of the simulated machine. This result suggested the following generalization: all models of machine are equivalent for the computational time (or the number of computational steps), provided we consider two models equivalent if the computational time of one of them is a polynomial in the computational time of the other. If this idea is correct, the exponential or polynomial character of a model is conserved when going from one model to another, hence the idea of defining the algorithmic complexity of a given problem from the number of computational steps $\mathcal{N}$ which are needed on a Turing machine to solve this problem. If $\mathcal{N}$ is a polynomial in n, then we say that the problem is "tractable", and if $\mathcal{N}$ grows faster than any polynomial in n, then the problem is termed "intractable". Adding or multiplying two integers is a "tractable" problem, while factoring is thought to be an "intractable" one, although there is no formal proof of this statement. Two important complexity classes are the **P** class, that of problems whose solution is "tractable", and the **NP** class, that of problems whose possible solutions can be *checked* in polynomial time. For example, it is "intractable" to factor an integer into primes, but once prime factors have been found, or guessed, it is "tractable" to check that the solution works, with a simple multiplication. Of course, $\mathbf{P} \subset \mathbf{NP}$, and there exists a famous conjecture $\mathbf{P} \neq \mathbf{NP}$, which has not been proved, or disproved, for the time being. Many complexity classes have been identified, by appealing in general to the computational model of Turing machines, but which are independent of this model provided we use a model which can be simulated, within a polynomial equivalence, by a Turing machine. One important complexity class is that of **NP**-complete problems, such as that of the traveling salesman: given a number of towns, the salesman must find an itinerary which goes once and only once through every town and minimizes the traveled distance. The discovery of a polynomial algorithm for any **NP**-complete problem would imply the existence of such an algorithm for all **NP** problems.

Up to now, we have only taken into account calculable problems. The *Church–Turing thesis*, which is universally accepted although it is by its very nature impossible to prove, states that *the class of functions which are calculable by a Turing machine corresponds exactly to the class of functions which we would naturally consider to be calculable using an algorithm.* There are properly identified problems for which it is known that there does not exist any algorithm, for example that of the halting of a Turing machine: the function which, to any program of a Turing machine, associates 0 or 1 according to whether the machine will stop or not. It is not a calculable function. Quantum computers do not seem to invalidate the Church–Turing thesis: functions which are calculable with a quantum computer are *a priori* the same as those calculable with a classical computer.

We already mentioned that the simulation of any model of a classical algorithm can be implemented on a Turing machine with an increase of computational time at most polynomial, and this result was postulated as a kind of "axiom" underlying the theory of algorithmic complexity. This is the strong version of the Church–Turing thesis: *any computational model can be simulated on a probabilistic Turing machine with an at most polynomial increase in the number of steps*. The importance of quantum computers is that they may invalidate this strong version: actually, if factoring is an "intractable" problem, which is suggested by experience but remains to be proven, then Shor's algorithm contradicts the strong version. With a quantum computer, it is possible to factor an integer into primes in a number of steps polynomial in n, while a classical computer must use an exponential one. Unfortunately, factoring is not a **NP**-complete problem, so that its polynomial solution does entail that of all **NP** ones.

8.4 Physical realizations

Concrete realizations of quantum computers are still in their infancy, and it is quite premature today to try to foretell what kind of device will be used for a quantum computer able to process thousands of qubits, if such a computer ever exists. Storage and processing of quantum information requires physical systems that obey the following conditions:

 (i) these systems must be scalable to a large number of qubits which are physically well-defined;
 (ii) it must be posssible to initialize all the qubits in the state $|0\rangle$;
(iii) the qubits must be carried by physical systems with a large enough lifetime, so that the coherence of quantum states is ensured during the entire calculation;
(iv) the systems must provide an ensemble of universal quantum gates: one-qubit gates and **cNOT** gates obtained by perfectly controlled manipulations;
 (v) there must exist an efficient procedure to measure the qubits at the end of the calculation (readout of the results).

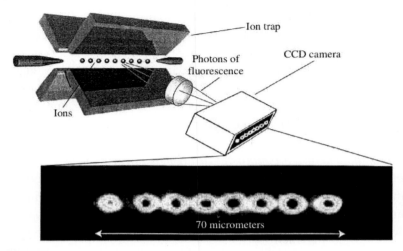

Figure 8.6.   Schematic description of a quantum computer using trapped ions. One distinguishes a row of trapped ions which are manipulated with a laser. Courtesy of Alain Aspect and Philippe Grangier, original figure from Rainer Blatt.

Public Enemy Number One of the quantum computer is its interaction with the environment, which leads to the decoherence phenomenon (Chapter 9) and the consequent loss of phase coherence in the superposition of quantum states. Calculations must be performed in a time shorter than the decoherence time τ_{dec}. If an elementary operation on a qubit, for example that of quantum gate, takes a typical time τ_{op}, the figure of merit of a quantum computer is the ratio

$$n_{op} = \frac{\tau_{dec}}{\tau_{op}}$$

which gives the maximum number of operations that can be performed.

There exist at present two main avenues for the effective realization of quantum computers. The first one uses systems where the qubits are carried by individual quantum particles: photons, atoms, ions.... These systems are dubbed "clean", as their control is relatively efficient, but they are not easily scalable. The second avenue uses qubits carried by collective systems, such as quantum dots or superconductors. The main strength of these systems is that they may take advantage of all the technology which has been developed in semiconductor physics for microchips and they are in principle more readily scalable, but they are "dirty", in the sense that the control of collective systems is less efficient than that of individual ones. The present state of the art is as follows: it has been possible to manipulate up to 10 qubits in perfectly controlled situations (seven in the case of Nuclear Magnetic Resonance (NMR)-based quantum computer, which, unfortunately, is not scalable), and in the absence of an unpredictable technological breakthrough, we shall probably have to wait for twenty or thirty years before a really efficient quantum computer sees the light of the day.

8.5 Further reading

There is no really elementary book about quantum computing. For an intermediate level presentation, the reader may consult Ekert [2006]. Introductory books are Le Bellac [2006], Mermin [2007] and Rieffel and Polack [2011]. A general overview is given by Scarani [2011], and one of the most promising implementation of quantum computers using trapped ions is described by Blatt [2004].

9

The Environment is Watching

It is generally assumed that the fundamental laws of physics are those of quantum physics, and that classical physics is only an approximation valid under certain conditions. However, the passage from quantum-to-classical and its conditions of validity are still hotly debated today. We may, for example, ask the following question: in a quantum world, particles exhibit interference phenomena. Why is it that interferences are never observed in a classical world except, of course, with classical waves? Equivalently, why is it that we do not see linear superpositions of macroscopically distinguishable states? A possible answer relies on the concept of *decoherence*. Although implicit in previous work, this concept was introduced explicitly at the beginning of the 1980s, and after a rather modest start, it has become more and more popular in domains like quantum information (Section 8.4) and quantum measurement (Section 10.3), where it plays a major role. After having explained the concept of decoherence with the help of an elementary example, we shall describe experiments where decoherence is manifest at *mesoscopic scales*, intermediate between microscopic and macroscopic scales, that is, on the order of a few μm.

9.1 Decoherence: An elementary example

To introduce the concept of decoherence, we use as an example the Mach–Zehnder interferometer of Section 1.2 with polarized photons. In that section, we took into account the spatial properties of the photons, that is their propagation in the interferometer, but we neglected their internal degree of freedom, their polarization, which is denoted P: $P = V$ for a polarization perpendicular to the interferometer plane, and $P = H$ for a polarization in this plane (Figure 9.1). We assume that photons enter the interferometer in a vertical polarization state V. As long as we use the interferometer with the setting of Figure 1.3, nothing is changed, and the polarization plays no role. However, an apparently minor modification will upend the picture drastically. Let us insert on the red path a birefringent plate Λ which transforms the vertical polarization into an horizontal one ($\theta = \pi/2$ in Figure 9.1). We shall need the general relations giving the action of a beamsplitter on an incident photon beam, which were already written down in another context in § 5.2.3. Assume that two amplitudes b_X and b_Y are incident on a beamsplitter along the directions OX and OY respectively, and let a_X and a_Y be the outgoing amplitudes along the same directions. The relation between amplitudes reads (Figure 9.2)

$$a_X = \frac{1}{\sqrt{2}}(b_X + b_Y), \quad a_Y = \frac{1}{\sqrt{2}}(b_X - b_Y). \tag{9.1}$$

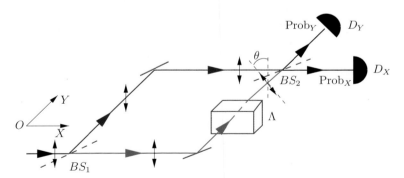

Figure 9.1. A "which way" experiment with polarized photons. Photons enter the Mach–Zehnder interferometer at the lower left side, and their polarization V is perpendicular to the horizontal plane of the figure. It is modified by the birefringent plate Λ (violet) on the red path, and makes an angle θ with the vertical axis. The discussion is carried out first in the case $\theta = \pi/2$, where the polarization in the red path is horizontal and acts as a marker of the path.

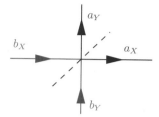

Figure 9.2. Action of a beamsplitter on two incident amplitudes b_X and b_Y. The outgoing amplitudes are a_X and a_Y.

Box 9.1. Action of a beamsplitter.

The minus sign in the equation giving a_Y in (9.1) can be understood from energy conservation in wave physics, or probability conservation in quantum physics. The intensity incident on the beamsplitter is

$$I_{inc} = b_X^2 + b_Y^2,$$

and the outgoing intensity is

$$I_{out} = a_X^2 + a_Y^2 = \frac{1}{2}(b_X + b_Y)^2 + \frac{1}{2}(b_X - b_Y)^2 = b_X^2 + b_Y^2 = I_{inc}.$$

The minus sign is clearly necessary if we wish to ensure $I_{out} = I_{inc}$, a relation which must hold true for a non-absorbing beamsplitter. To be mathematically rigorous, we should use complex numbers and write $|a_X + a_Y|^2$, not $(a_X + a_Y)^2$.

In the situation described in Section 1.2, the amplitudes incident on the second beamsplitter BS_2 were $b_X = b_Y = a/\sqrt{2}$ (Figure 1.3), whence $a_X = a$, $a_Y = 0$: all the photons leave the second beamsplitter along the horizontal path. We must now take into account the polarization and introduce four different amplitudes instead of two. In order to compute them, we introduce four amplitudes incident on the beamsplitter, which now depends of the polarizations V or H

$$b_X(V) = \frac{a}{\sqrt{2}}, \quad b_X(H) = 0, \quad b_Y(V) = 0, \quad b_Y(H) = \frac{a}{\sqrt{2}}$$

and use equation (9.1), which gives the amplitudes leaving BS_2.

1. Photon leaving in a direction parallel to OX and vertically polarized: probability amplitude $a_X(V) = a/2$.
2. Photon leaving in a direction parallel to OX and horizontally polarized: probability amplitude $a_X(H) = a/2$.

3. Photon leaving in a direction parallel to OY and vertically polarized: probability amplitude $a_Y(V) = a/2$.
4. Photon leaving in a direction parallel to OY and horizontally polarized: probability amplitude $a_Y(H) = -a/2$.

In order to compute probabilities, we appeal to rule number 3 of § 1.6.2, taking into account that the polarization states V and H are *distinguishable*. We could indeed analyze the photon polarization after the second beam-splitter thanks to a PBS, and identify the polarization. We are left with four different probabilities, $\text{Prob}_X(V), \ldots, \text{Prob}_Y(H)$, with the same notations as for the amplitudes. Squaring the amplitudes, we see that all four probabilities are equal to $1/4$ $(=(1/2)^2 = (-1/2)^2)$

$$\text{Prob}_X(V) = \text{Prob}_X(H) = \text{Prob}_Y(V) = \text{Prob}_Y(H) = \frac{1}{4}.$$

The probability Prob_X that a photon leaves the second beamsplitter along OX is obtained by adding the contributions from the vertical and horizontal polarizations

$$\text{Prob}_X = \text{Prob}_X(V) + \text{Prob}_X(H) = \frac{1}{2}.$$

Interference has disappeared! We have a concrete realization of what is called a "which way experiment", where the photon path has been determined: *the polarization acts as a marker of the path followed by the photon*, since observing a polarization V (H) means that the photon followed the blue (red) path. Then we can divide the photons into two groups, those which followed the blue path and those which followed the red one. As explained in Section 1.4, no interference can occur in any of these groups, and this property holds globally. We now add three *fundamental* remarks.

1. There was no possible perturbation of the spatial properties of the photons. That interference is destroyed cannot be blamed on some "uncontrollable perturbation" due to measurement. It is the labeling of paths which is at the basis of interference destruction.
2. It is not necessary to observe the polarization for the interference to be destroyed. It is enough that the experimental device can, in principle, give access to this information. The photon behavior is not linked to the

fact that an observation is performed, or not, *but to the configuration of the experimental device*, a property already observed in the case of the delayed choice experiment (Section 1.7).

3. By manipulating the photon just after the second beamsplitter, we could erase the information contained in the polarization and recover interference. It is thanks to an analogous, but slightly different, device that we can perform the delayed choice experiment. *Until a detector has been triggered*, we cannot claim positively that the photon has followed one of the two paths. As J. A. Wheeler put it: "No elementary phenomenon is a phenomenon until it is a registered phenomenon."

We conclude this Section with a slight generalization of our argument, by allowing the birefringent plate Λ to rotate the polarization by an angle θ, instead of $\pi/2$. We obtain immediately the four amplitudes

$$a_X(V) = \frac{a}{2}(1 + \cos\theta), \quad a_X(H) = \frac{a}{2}\sin\theta$$

$$a_Y(V) = \frac{a}{2}(1 - \cos\theta), \quad a_Y(H) = -\frac{a}{2}\sin\theta,$$

which leads to the probabilities

$$\text{Prob}_X = \frac{1}{4}(1 + \cos\theta)^2 + \frac{1}{4}\sin^2\theta = \frac{1}{2}(1 + \cos\theta)$$

$$\text{Prob}_Y = \frac{1}{4}(1 - \cos\theta)^2 + \frac{1}{4}\sin^2\theta = \frac{1}{2}(1 - \cos\theta),$$

with, of course, $\text{Prob}_X + \text{Prob}_Y = 1$. The interference visibility $\mathcal{V}$ is defined by

$$\mathcal{V} = \frac{\text{Prob}_X - \text{Prob}_Y}{\text{Prob}_X + \text{Prob}_Y} = \cos\theta. \tag{9.2}$$

By combining the OX and OY paths, the second beamsplitter allows us to test the *coherence* between the two propagation states. The coherence is maximum ($\mathcal{V} = 1$) for $\theta = 0$, and it vanishes ($\mathcal{V} = 0$) for $\theta = \pi/2$. In other words, the beamsplitter allows us to check the superposition of the two paths. This superposition is coherent (or linear) for $\theta = 0$ ($P = V$) and incoherent for $\theta = \pi/2$ ($P = H$).

The Mach–Zehnder interferometer with polarized photons illustrates a case where there exists a quantum correlation between a spatial degree of freedom, the propagation along *OX* or *OY*, and a polarization degree of freedom. This is another example of entanglement, and this entanglement of a spatial degree of freedom with a polarization (internal) degree of freedom has allowed us to influence the coherence of the first one. If we observe only the spatial degree of freedom, then we notice a progressive loss of coherence as the angle θ increases from 0 to $\pi/2$. The loss of coherence is complete when the internal degrees of freedom are fully distinguishable, for example, the states V and H. This loss of coherence due to entanglement is precisely what is called *decoherence*.

The interference experiments have often been discussed in the framework of *Bohr's complementarity principle*. If the two paths are distinguishable, there is no interference and we observe the particle aspect of quantum particles. If they are not distinguishable, we observe interference, and thus the wave aspect. The two aspects are claimed to be mutually exclusive. However, interference experiments reveal both aspects, because the particles are localized when they are detected and the interference pattern is obtained in a discrete-like manner. Moreover, by varying the angle θ, we may go continuously from the wave aspect to the particle aspect. What is really important is the entanglement between the degree of freedom we are directly interested in and another one, which gives access, (or not, or maybe only in part) to information on the path.

9.2 Environment and decoherence

In what follows, it will be convenient to introduce Dirac's notation in order to make explicit use of quantum states, rather than probability amplitudes. The two descriptions are physically equivalent, but Dirac's notation is often more transparent. With this notation, we represent a quantum state $\bullet$ by $|\bullet\rangle$: it is the *state vector* of the quantum system, see the preceding chapter for qubit states and Appendixes A.2.1 and A.4.1 for more details. Let us give an example: the quantum state of a vertically polarized photon is written $|V\rangle$, that of an horizontally polarized photon $|H\rangle$. The fundamental property of state vectors is that they can be added, much as probability amplitudes: they obey the superposition principle, and this is the reason behind the physical equivalence of the two approaches. A linear polarization state $|P\rangle$ making an angle θ with the vertical axis is written in the form of a state vector as

$$|P\rangle = \cos\theta\,|V\rangle + \sin\theta\,|H\rangle. \tag{9.3}$$

This equation implies that the probability amplitude of finding the photon in the state $|V\rangle$ ($|H\rangle$) is $\cos\theta$ ($\sin\theta$), and equation (9.3) is the mathematical translation of a linear (or coherent) superposition such as introduced in § 1.3.3: the state $|P\rangle$ is a linear superposition of states $|V\rangle$ and $|H\rangle$. It is important to observe that the factor $\cos\theta$ in (9.3) is precisely that which fixed the coherence (9.2) in the preceding section. This factor measures the overlap of the states $|P\rangle$ and $|V\rangle$, and is mathematically equal to their scalar product $\langle P|V\rangle$. With Dirac's notation, the states of propagation along the directions OX and OY are written $|X\rangle$ and $|Y\rangle$, and the spatial part of the state vector just before the second beamsplitter is

$$|\psi\rangle = \frac{1}{\sqrt{2}}(|X\rangle + |Y\rangle). \tag{9.4}$$

The mechanism described in the preceding section contains the essentials of the decoherence phenomenon, considered as a consequence of entanglement, *independently of the mechanism used to generate it.* The usual mechanism to generate entanglement is the interaction of the quantum system S we are interested in with an environment $\mathcal{E}$. Let us assume that S is initially in a superposition of two fully distinguishable, that is, mathematically orthogonal states, $|S_1\rangle$ and $|S_2\rangle$

$$|\psi\rangle = \frac{1}{\sqrt{2}}(|S_1\rangle + |S_2\rangle), \tag{9.5}$$

where $|S_1\rangle$ and $|S_2\rangle$ are the analogues of the spatial states $|X\rangle$ and $|Y\rangle$ of (9.4). This choice is made for easy comparison with the interferometer case. In the general case, we would have $|\psi\rangle = \alpha\,|S_1\rangle + \beta\,|S_2\rangle$ with $|\alpha|^2 + |\beta|^2 = 1$. Because of the interaction with the environment, time evolution entangles the system with the environment, for example, $|E_1\rangle$ for $|S_1\rangle$ and $|E\rangle$ for $|S_2\rangle$, with

$$|E\rangle = \cos\theta\,|E_1\rangle + \sin\theta\,|E_2\rangle,$$

where the states $|E_1\rangle$ and $|E_2\rangle$ are fully distinguishable. The system–environment interaction then generates an entangled state

$$|\Psi_{AE}\rangle = \frac{1}{\sqrt{2}}(|S_1\rangle|E_1\rangle + |S_2\rangle|E\rangle), \tag{9.6}$$

where we note the correspondence with the interferometer case

$$|X\rangle \implies |S_1\rangle, \quad |Y\rangle \implies |S_2\rangle, \quad |V\rangle \implies |E_1\rangle, \quad |P\rangle \implies |E\rangle.$$

Then, for the same reason as in the preceding Section, the initial coherence contained in (9.5) is reduced by the factor $\cos\theta$, which is nothing other than the scalar product $\langle E_1|E\rangle$. In practice, the system S interacts with a complex environment with a large number of degrees of freedom. As a consequence, the $\cos\theta$ factor decreases with a characteristic time τ_{dec}, the *decoherence time*: the dynamics introduces a phase factor which is different for each degree of freedom. The scalar product $\langle E_1|E\rangle$ is a sum of a large number of sine functions oscillating with different frequencies and it tends rapidly to zero for large times. This qualitative argument is confirmed by semi-realistic models of $S - \mathcal{E}$ interactions. The decoherence time may vary from a fraction of a second (experiments of Serge Haroche and collaborators at the ENS, Paris) to 10^{-17} s for a heavy molecule prepared as a linear superposition of two states localized at a distance of 1 nm in the air. When the states $|E_1\rangle$ and $|E\rangle$ have become orthogonal ($\theta = \pi/2$), no measurement performed on the system S *alone* will be able to display a coherence between the states $|S_1\rangle$ and $|S_2\rangle$. Much as the polarization acts as a marker of the photon path, the state of the environment acts as marker of that of S. It must be understood that the origin of decoherence does not lie in the complex, chaotic, or uncontrollable character of the environment, even though these characteristics are quantitatively important to determine the decoherence time, but rather in the $S - \mathcal{E}$ interactions. One should not confuse decoherence, *which is a quantum phenomenon*, with noisy interactions. In addition to the decoherence time, environmental decoherence exhibits a second specific property with respect to the interferometer case. If S interacts with particles of $\mathcal{E}$ which then travel far away from the interaction zone, the $S - \mathcal{E}$ correlations are delocalized, as in the case of Bell's correlations. This makes environmental decoherence irreversible in practice, because it is impossible to send all the particles back to the interaction region.

Two important points should be noted. First, decoherence and irreversibility are linked, but they are not identical. Let us take the case of a heavy molecule propagating in a dilute gas of light molecules, which is initially in a superposition of states localized at two different points in space. Because of collisions with the light molecules, the heavy molecule loses energy, and the time characteristic of this energy loss is τ_{irr}. However, $\tau_{irr} \gg \tau_{dec}$, since a large number of collisions is needed to slow down the molecule, but a single collision is enough to destroy coherence. Second, we should not think that the environment perturbs the system; on the contrary, *it is the system which perturbs the environment*. As can be seen in equation (9.6), the system has not changed, but the environment has been modified.

The destruction of interference due to interactions with the environment has been checked in a number experiments. One with the easiest interpretation was conducted by Anton Zeilinger and collaborators in Vienna, who performed Young slit type experiments with heavy molecules of the fullerene kind, whose molecular weight is on the order of 1500. It is of course more and more difficult to observe interference when the masses of the particles increases, as the distance between fringes varies as λ_{dB}/d, where λ_{dB} is the de Broglie wavelength of the molecules and d the distance between the slits. Since $\lambda_{dB} = h/mv$, where m and v are the mass and velocity of the molecules, the distance between fringes decreases as the

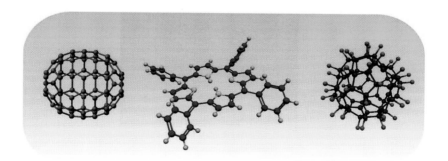

Figure 9.3. Molecules used in the experiments performed by Anton Zeilinger and collaborators: (a) C_{70}, (b) $C_{44}H_{30}N_4$, (c) $C_{60}F_{48}$. From Arndt *et al.* [2005], courtesy of Anton Zeilinger.

inverse of the mass. It is possible to alleviate in part this difficulty by using a Talbot–Lau interferometer, where this distance decreases only as $1/\sqrt{m}$. However, there is another reason which makes the observation of interference more difficult the heavier the molecules, and it is precisely their interaction with the environment. These heavy molecules possess hundreds of degrees of freedom, which are only asking to interact with the environment. These interactions are mainly photon emission and collisions with the light molecules of the residual gas. We obtain either entangled photon–molecule states, or entangled heavy molecule–gas molecules states. When traveling from the source to the observation screen, molecules leave an imprint in the environment. This imprint contains information on the path followed in the interferometer, and thus on the chosen slit. Experiments (Figure 9.4)

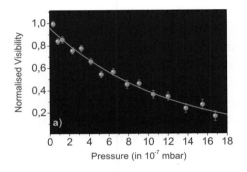

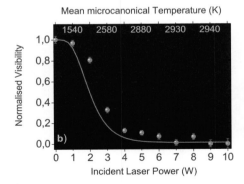

Figure 9.4. Blurring of interferences due to collisions and photon emission. The vertical axis gives the visibility (9.2) of the interference fringes. Upper figure: influence of the pressure. Lower figure: influence of the temperature, linked to the intensity of the laser which is initially shone on the molecule. From Arndt *et al.* [2005], courtesy of Anton Zeilinger.

confirm the progressive destruction of interference when the temperature of the molecules rises, or when the residual gas pressure increases. In the first case, emission of photons of shorter and shorter wavelengths owing to the excitation of higher levels of the molecules leads to a better resolution of the path: if the emitted wavelength is on the order of d, we may, in principle, distinguish between the slits. In the latter case, the Brownian motion of the molecules induced by collisions with the gas particles is responsible for the blurring of interference. In both cases, it is possible to calculate theoretically the deterioration of the fringe visibility, and the calculations are in good agreement with experiment. There is little doubt that decoherence is at work in this experiment. Another equivalent way of looking at the problem is to show that the two processes allow us to partially localize the molecules, and when they can be perfectly localized, no interference can survive: the molecule path is fully known.

A remarkable property of decoherence is that the decoherence time decreases very fast when the size of the observed object grows, because the number of degrees of freedom able to interact with the environment grows rapidly with the size. Then, for a macroscopic and even mesoscopic object, this time is so short that decoherence is not observable in practice, except in experiments like those conducted by the group of Serge Haroche at the Paris ENS, which were able to catch decoherence in the act. This property is confirmed by semi-realistic models, where the decoherence time decreases as the inverse of the size squared: linear superpositions of macroscopically distinguishable states are almost immediately destroyed. What must be understood is that an object is not "intrinsically quantum" or "intrinsically classical". According to its environment, the same object, for example a heavy molecule, will display a quantum behavior (interference) or a classical one (no interference). Macroscopic and even mesoscopic objects exhibit a classical behavior because they possess an enormous number of degrees of freedom interacting with their environment, and the decoherence time is so short that it is not accessible in practice.

However, theory does not predict any upper limit of size or of complexity for a quantum object, a limit beyond which quantum effects (superposition or coherence) would be absent, and we may well imagine observing in the future interference with proteins or viruses, provided experimentalists are able to strictly control the environment. Superconducting circuits have

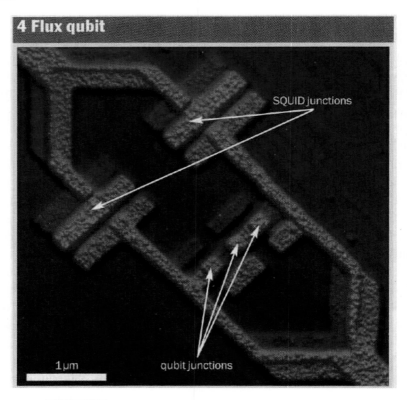

Figure 9.5. A superconducting qubit. Courtesy of Hans Mooij.

recently been built, whose (mesoscopic) size is on the order of 1 μm, and where an electric current may be in a linear superposition of two currents flowing in opposite directions: such a state is an example of a superposition of macroscopically distinguishable states, as the number of charge carriers may reach 10^9 (Figure 9.5). The superconducting qubits are genuine atoms of mesoscopic size and have an energy spectrum much as an ordinary atom. Other mesoscopic devices have been thought of, and some of them will certainly be implemented in the near future: nano-mechanichal devices with a discrete energy spectrum, or entanglement of a photon and a mirror with $\sim 10^{15}$ atoms. To conclude, the present state of quantum theory does not allow us to draw in natural way a frontier between a quantum and a classical world.

9.3 Further reading

Zurek [1991] and its updated version [2003] may serve as an introduction. Important experiments on decoherence were performed by Brune *et al.* [1996] and Arndt *et al.* [2005]. At an advanced level, the book by Haroche and Raimond [2006] describes many aspects of decoherence, while Schlossauer's [2007] is fully devoted to this topic.

10

Interpretations

Although nobody can question the practical efficiency of quantum mechanics, there remains the serious question of its interpretation. As Valerio Scarani puts it,

> "We do not feel at ease with the indistinguishability principle (that is, the superposition principle) and some of its consequences."

Indeed, this principle which pervades the quantum world is in stark contradiction with our everyday experience. From the very beginning of quantum mechanics, a number of physicists — but not the majority of them! — have asked the question of its "interpretation". One may simply deny that there is a problem: according to proponents of the minimalist interpretation, quantum mechanics is self-sufficient and needs no interpretation. The point of view held by a majority of physicists, that of the Copenhagen interpretation, will be examined in Section 10.1. The crux of the problem lies in the status of the state vector introduced in the preceding chapter to describe a quantum system, which is no more than a symbolic representation for the Copenhagen school of thought. Conversely, one may try to attribute some "external reality" to this state vector, that is, a correspondence between the mathematical description and the physical reality. In this latter case, it is the measurement problem which is brought to the fore. In 1932, von Neumann was first to propose a global approach, in an attempt to build a purely quantum theory of measurement examined in Section 10.2. This theory still underlies modern approaches, among them those grounded on decoherence theory, or on the macroscopic character of the measuring apparatus: see Section 10.3. Finally, there are non-standard interpretations such as Everett's many worlds theory or the hidden variables theory of de Broglie and Bohm (Section 10.4). Note, however, that this variety of interpretations has no bearing whatsoever on the *practical* use of quantum mechanics. There is no controversy on the way we should use quantum mechanics!

Box 10.1. External reality.

From a philosophical point of view, the concept of "reality" needs to be considered with the utmost caution. As my own philosophical competencies are close to zero, I won't venture onto this ground. The "reality" I am referring to is nothing other than the naïve reality of physicists: there is an external world which exists independently of the description which we may give of it, or of the observations which we may perform on it.

10.1 The Copenhagen interpretation

In classical physics, the state of a physical system is essentially a list of its properties: position and velocity for a point particle, amplitudes of the electric and magnetic fields for an electromagnetic field, and so forth. These properties are collections of numbers. Dynamics specifies how they evolve in time according to a deterministic law: given the initial conditions and the equations of motion, there is no ambiguity on the future evolution. It must be understood that this is a statement of principle: the uncertainties on the initial conditions, the difficulties in solving the equations of motion and the possible chaotic character of these equations limit considerably its practical value but, in principle, we may visualize the space time evolution and there is no difference between the list of its properties and the reality attached to a physical system. On the contrary, quantum theory allows us, given initial conditions, to predict only the *probability* of observing such or such result. Moreover, the experimental violation of locality for some quantum systems shows that we cannot visualize their continuous evolution in space-time. Given these two characteristics of quantum physics, we are led to ask the question of the status of the state vector. Is it only a catalog which registers the manipulations inflicted on the system by an experimentalist, often called an "observer" in this context, or does the state vector describe an external reality to which it is attached? This choice lies at the heart of the interpretation problems of quantum mechanics.

Is it necessary, useful, or completely superfluous to be interested in interpretation problems? The minimalist interpretation clearly chooses the last option: quantum mechanics is self-sufficient. According to Christopher Fuchs and Asher Peres, "Quantum mechanics needs no interpretation". These authors claim that the only thing we must require from the theory is that it calculates correctly probabilities within the general scheme of an experiment performed with quantum objects, a scheme which we sum up briefly: in a first stage, a preparation procedure allows us to obtain an ensemble of quantum objects all in the same state. This preparation stage is followed by a quantum evolution and, after some time, we perform a measurement on the final state. We then repeat several measurements on this series of quantum objects all prepared in exactly identical conditions. Let us illustrate this scheme with photon polarization: in the preparation stage, a polarizer such as that of Figure 2.1 selects photons, all in a vertical polarization state. After manipulation of these photons, we analyze their final polarization with a PBS and two detectors (Figure 2.2). In more complex cases, we may use a "multiple channel PBS" (see Figure 11.7). In principle, the theory allows us compute the probability $\text{Prob}(D_i)$ that detector D_i in Figure 11.7 be triggered and, if we have prepared an ensemble of N identical initial states, detector D_i will be triggered a number of times $N \times \text{Prob}(D_i)$ on average. It is important to observe, for example in the case of Figure 2.2, that a measurement is completed *only* when a detector has registered this particular photon, and *not* as soon as the photon has taken one of the two paths at the PBS exit. We recall once more the warning of J.-A. Wheeler: "No elementary phenomenon is a phenomenon until it is a registered phenomenon." The relevance of this statement is particularly clear in the delayed choice experiment of Section 1.7, which was actually proposed by Wheeler!

In the minimalist interpretation, the concepts of quantum theory — probability amplitudes, state vectors, and so forth — are mere calculational tools and do not represent any external reality. The state vector represents nothing other than a preparation procedure. Quantum theory does not apply to individual systems, but only to ensembles. This point of view, sometimes attributed to Einstein, who considered anyway quantum theory as incomplete, was quite natural when experimentalists could only manipulate large numbers of quantum particles, for example billions of sodium atoms in order to measure their spectrum, and the probabilistic character could be reasonably linked to the necessity of experimenting on large ensembles. Nowadays, this point of view may be harder to defend: how can we avoid considering the individual state of a single trapped ion which we can observe for hours, making transitions between different energy levels? However this argument, even if it shows that the minimalist interpretation is not very natural in this context, does not allow us to reject it definitely. Indeed, we can always interpret a series of observations on a single ion as a series of observations performed on identical quantum states.

The Copenhagen interpretation, devised between 1925 and 1927 mainly by Bohr and Heisenberg, applies quantum theory to individual systems, and not to ensembles. It was initially grounded on two pillars: Heisenberg's inequalities (§ 4.3) and *complementarity*. In its most elementary version, complementarity states that we cannot in an experiment observe at the same time the particle and the wave aspect of a quantum particle: these two aspects are termed complementary. However, (i) the experiments described in Figures 1.8 and 1.9 display both aspects, because photon or atom detection is localized and creates an interference pattern in a discrete-like manner and (ii) the device of Section 9.1 shows that one can go continuously from the particle aspect (absence of interference) to the wave aspect (interference), thanks to entanglement between the spatial and internal degrees of freedom. Although complementarity is still mentioned in recent textbooks, it seems that, at least in this elementary form, this concept is no longer useful today: see Kaiser *et al.* [2012]. Finally, according to Bohr and Heisenberg, the probabilistic character originates in that, when a measurement is performed, the quantum object interacts with a macroscopic device which perturbs it in an uncontrollable way. This is the famous "uncontrollable perturbation" of the measurement process, which is also still quoted in a majority of textbooks. The measurement results must be expressed in classical terms, which is the only way to communicate to other people what we have done. Bohr and Heisenberg disagree on the role of the observer, to whom Heisenberg, contrary to Bohr, attributes a central role.

Box 10.2. Copenhagen interpretation.

There is no canonical version of this interpretation. Bohr's point view changed with time and differs considerably on certain points from that of Heisenberg. It looks like each physicist interested in the subject develops his own version of the Copenhagen interpretation. Entering into more details would necessitate hundreds of pages.

In 1935, the EPR article (Chapter 3) led Bohr to clarify, and even revise his position in depth. Indeed, EPR underlined that one could access a physical property of a quantum object without directly interacting with it. In the measurement process, the perturbation did not seem to play a fundamental role. This absence of a perturbation is also emphasized in the example of Section 9.1, where the photon path is determined thanks to its polarization, without any perturbation whatsoever of the path. Many other examples of measurement without perturbation, also called non-destructive measurements, have been proposed, and some of them have been actually realized. This point is not questioned by Bohr in his answer to the EPR article, which takes as his main targets the rather fuzzy notion of "element of reality" introduced by EPR and their claim of the incomplete character of quantum mechanics. Finally, as already mentioned, Bohr takes advantage of his answer to clarify his interpretation of quantum mechanics.

Bohr's article is difficult to read, and I tried to reorganize his arguments, hoping not to have distorted (too much) his line of thought. First of all, Bohr asserts once more that the apparatus which allows us to experiment on quantum systems must be *described* in classical terms, because only a vocabulary using terms defined in classical physics allows us to communicate to other persons what we have done. For example, we may communicate the location of the detectors which registered photons and the time when they were triggered. This does not necessarily imply that the apparatus itself operates within the laws of classical physics; this is a delicate point still under debate. We have already encountered Bohr's second point when discussing the minimalist interpretation: a quantum phenomenon can be considered as such only when the experiment is completed and when its results have been registered thanks to irreversible processes.

Box 10.3. Elements of reality.

According to EPR, if we are able to predict the value of a physical property attached to a physical system without any direct interaction whatsoever with this system, then there exists an element of reality attached to this physical property, which must find its place in the theoretical description. This EPR concept relies implicitly on separability, that is, the possibility of ascribing properties to individual components of a composite system, for example, the two photons of Chapter 3, when the two subsystems are too far away to be able to interact and to communicate.

It follows, and this is the third point, that the properties of a quantum system, and the notion of "physical reality" attached to it, cannot be dissociated from the configuration of the experimental apparatus which is used for its observation. This third point lies at the heart of Bohr's answer: the notion of "element of reality" introduced by EPR is not part of the framework of quantum theory, because the experimental conditions are an intrinsic part of the description of a quantum system to which we may ascribe the notion of "physical reality", and one cannot act as if this reality had an existence independent of these conditions. For Bohr, there is no such thing as intrinsic properties of a quantum system, and its representation by a state vector can only be a symbolic one, which does not necessarily correspond to a physical reality. In order to illustrate this argument, let us return to Figure 3.3, where Alice and Bob can, for example, use for their analysis the configuration (or basis, see § 2.3.1) $\{VH\}$ or the incompatible configuration $\{DA\}$. The two configurations are mutually exclusive and, according to Bohr, the two photons traveling toward Alice and Bob from the source S simply do not correspond to the same physical reality: there does not exist any physical reality corresponding to the two-photon system between the source and the detectors.

Finally, and this is the fourth point, the concept of complementarity is grounded in the notion of mutually exclusive configurations. The two experimental configurations we just mentioned, $\{VH\}$ and $\{DA\}$, are mutually exclusive. The $\{VH\}$ and $\{DA\}$ polarized photons are linked to two mutually exclusive experimental configurations. In any concrete situation, mutually exclusive experimental configurations allow us to describe completely and without any ambiguity all quantum phenomena. When defined as above, complementarity lies at the heart of the exhaustive description of quantum phenomena, and the criticism by EPR of the incomplete character of this description has no basis. It is worth adding that there is no problem with locality either, because it is only *after* they have completed their respective measurements that Alice and Bob may decide to meet in order to compare their results, and to conclude that they are correlated. It is only the formalism which is non-local.

Bohr's point of view draws a line between the system to be studied and the experimental apparatus. According to Bohr, this is the main difference between quantum and classical physics because, in the latter case, this distinction does not exist. The description of a classical system is intrinsic: it does not depend on the experimental environment. It is not necessary that the quantum system be microscopic and the apparatus macroscopic, and the system/apparatus dividing line should not be confused with a microscopic/macroscopic distinction, or even with a classical/quantum frontier. However, the necessity of drawing this line cannot be easily formalized. As we saw in the preceding chapter, the classical limit of quantum physics is far from well understood. And, at the end of the day, we cannot bypass the observation that the apparatus itself is governed by the laws of quantum physics.

Be that as it may, the two key points of the Copenhagen interpretation may be summarized as follows.

1. It is the experimental device used for observation which allows us to define a quantum system. This device must be described in classical terms.

2. The concepts introduced in quantum physics — probability amplitudes, state vectors and so forth — are only calculational tools, which do not allow us to visualize quantum evolution in space-time between preparation and measurement. One should not even attempt to use such a visualization.

A variant of Bohr's point of view, which is quite popular today, is based on the notion of information: the state vector does not describe the reality of a quantum system, but only the *information* we have on it. One strong argument in favor of this interpretation is grounded in the analysis of Section 3.4, which we reformulate using Dirac's notation for state vectors. Let us start from a polarization entangled state vector $|\Phi\rangle$

$$|\Phi\rangle = \frac{1}{\sqrt{2}} (|V_A\rangle|V_B\rangle + |H_A\rangle|H_B\rangle), \qquad (10.1)$$

where $|V\rangle$ and $|H\rangle$ are states with vertical and horizontal polarizations, respectively, whereas A and B label Alice and Bob. Let us assume that Alice performs a measurement of her photon $\mathcal{A}$ in the configuration $\{VH\}$, with the result $|V_A\rangle$. The state vector after this measurement is no longer $|\Phi\rangle$ (10.1), but $|\Phi_V\rangle = |V_A\rangle|V_B\rangle$, a process which is called *state vector collapse*, and it means that the photon arriving at Bob is vertically polarized, while before the measurement it had no well-defined polarization. Even if Alice and Bob are far away from each other and cannot communicate, the passage from $|\Phi\rangle$ to $|\Phi_V\rangle$ looks instantaneous. Yet, in the absence of superluminal communication, it is impossible that Alice's measurement could influence Bob's photon, considered as a physical object, while the *information* we have on this state has been modified. Before Alice's measurement, our information is that Bob's photon has no well-defined polarization, meaning that all polarization states are equally probable, a situation which is not described by a state vector, but by a density matrix. After Alice's measurement, Bob's photon is vertically polarized, and Alice is perfectly aware of it, but Bob has no way of obtaining this information, and his mathematical description of the photon is unchanged. The physical reality of Bob's photon is unchanged, but Alice and Bob have different information. The state vector is not attached to the physical reality of the photon, but to the information we have on it. It follows that the notion of an "unknown quantum state" is an oxymoron: by definition, a quantum state must be known to someone.

10.2 von Neumann's theory

In addition to the elusive character of the apparatus/system frontier, we may find another drawback in Bohr's interpretation: it does not correspond to the usual practice of those physicists, theorists as well as experimentalists, who make use of quantum physics everyday. Indeed, in their actual practice, physicists never question, whether openly or implicitly, the correspondence between the state vector and the reality of the physical system under study. Bohr's interpretation, which always refers to an experimental device to define a quantum system, is in practice a cumbersome framework to work with, and physicists give spontaneously an existence to a quantum system, independent of its observation. For example, an optician sending a single photon in a fiber will have no doubt that a photon "propagates" in the fiber, although, strictly speaking, he or she can only check the photon emission from the source and its detection at the the fiber exit. Similarly, a LHC physicist won't admit easily that a proton circulating in the ring has no intrinsic reality from its injection in the accelerator to the detection of the collision products by the detectors. But, if we adopt the "realist" point of view of the state vector, then we must face the quantum measurement problem.

Bohr was not preoccupied with the details of the measurement; he only had to make sure of two of its characteristics:

1. An *amplification* allowing one to translate a microscopic phenomenon into a macroscopic result, traditionally the position of a pointer on a dial.
2. An *irreversible* process allowing the result to be registered once for all, without any possibility of evolution backward in time. Let us recall the example of the PBS in the preceding section: as long as the photon has not triggered a detector, no measurement has been performed.

In 1932, von Neumann published a seminal paper, where he laid the foundations of a fully quantum theory of measurement. He observes that there are two kinds of evolution in quantum physics.

(i) An evolution of the deterministic kind, governed by Schrödinger's equation. Once the preparation phase is completed, a quantum state evolves in an entirely deterministic way governed by the initial conditions and the interactions, much as the evolution of a classical particle is determined by the initial conditions, positions and velocities, and by the forces. This evolution is also *reversible*: one may go backward in time. Given the state vector at time t, we may revert to the state vector at a time $t_0 < t$.

(ii) An evolution which is irreversible and probabilistic when a measurement is performed. In the preceding section, we witnessed the state vector collapse $|\Phi\rangle \rightarrow |\Phi_V\rangle = |V_A\rangle|V_B\rangle$. This evolution is probabilistic, because with a 50% probability, we could as well have observed the evolution $|\Phi\rangle \rightarrow |\Phi_H\rangle = |H_A\rangle|H_B\rangle$. This evolution is also irreversible, as we cannot recover the initial state $|\Phi\rangle$ neither from $|\Phi_V\rangle$ nor from $|\Phi_H\rangle$.

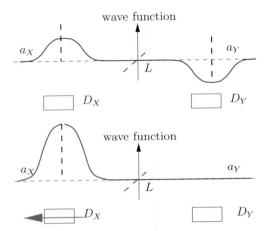

Figure 10.1. Artist's view of wave function collapse. We start from the scheme of Figure 1.4, where single photons are incident on a balanced beamsplitter. Before detection, the probability amplitude (or the wave function) is given by the upper figure. When detector D_X has performed a non-destructive measurement by registering the passage of a photon without absorbing it (easier said than done!), the wave function is transformed: lower figure. One must not think that this description corresponds to something which "really" happens in a laboratory!

This latter evolution is called, as already mentioned, state vector (or wave function) collapse (Figure 10.1). However, this collapse relies on an implicit technical condition: the measurement must be non-destructive, that is, the quantum particle must not be destroyed by the measurement, in which case knowledge of the state vector would be pointless. Yet, in practice, most of the measurements are destructive ones. For example, if we come back to Figure 1.4, a single photon triggers one of the two detectors with a 50% probability, but it will be absorbed and not available for future experiments. This is the reason why we have assumed in Figure 10.1 a non-destructive measurement. It is possible today to perform such measurements in sophisticated experiments, for example, to detect the presence of a photon without destroying it, and thus to find situations where state vector function collapse is a relevant concept. However, it can be shown that state vector collapse is fully equivalent to Born's rule (rule 2 of § 1.6.2 and Appendix A.4.1): if we perform successive measurements, correlations between these measurements may be computed either from Born's and Bayes's rules (Appendix A.3.1), or from state vector collapse, with identical results. State vector collapse is nothing other than a convenient reformulation of the basic rules. Moreover, for the proponents of the thesis "state vector = information", it is clear that state vector collapse can in no way be a physical process which takes place in a laboratory, but is nothing other than the modification of the information we have on the state. The version of quantum mechanics relying on the evolution postulates (i) and (ii) is that which is usually found in textbooks, and is often called *the standard*, or *orthodox, interpretation*.

Box 10.4. Wave function collapse and Bohr's interpretation.

In contradiction to a widely held view, wave function collapse is not part of the Copenhagen interpretation, at least in Bohr's version. Wave function collapse has clearly no meaning at all in the minimalist interpretation, which relies on ensembles. The (excellent) textbooks by Leslie Ballentine and by Asher Peres are two remarkable exceptions which avoid the textbook orthodoxy.

Resorting to two kinds of evolution is not very satisfactory: indeed, if quantum theory is universally valid, the system under study and the measuring apparatus are both governed by an evolution of type (i). Yet, the contact between the apparatus and the system is translated into a probabilistic evolution of type (ii), which may be considered as an internal inconsistency of the whole scheme. For this reason, von Neumann proposed to extend as far as possible the deterministic evolution of type (i). He assumed that the system to be measured S is subject to a quantum interaction with the apparatus A. Initially, the system is in a state $|S_i\rangle$ and the apparatus in a neutral state $|A_0\rangle$: the global initial state is then the product $|\Phi_i\rangle = |S_i\rangle|A_0\rangle$. A quantum interaction transforms this initial state $|\Phi_i\rangle$ into $|\Psi_i\rangle = |S_i\rangle|A_i\rangle$, where the states $|A_i\rangle$ and $|A_j\rangle$ may be distinguished without any ambiguity when $i \neq j$. Mathematically, this means that the two states are orthogonal: $\langle A_i|A_j\rangle = \delta_{ij}$. The apparatus is then a reliable pointer for a measurement of S: there is a one-to-one correspondence between the states of S and those of A. The state $|A_i\rangle$ of the apparatus may be identified with the position of a pointer on a dial: the pointer indicates the "direction" $|A_i\rangle$, and reading the dial reveals the state $|S_i\rangle$ of S. Up to now, everything seems to be going smoothly, but there is a hitch: assume S to be in a linear superposition $|S\rangle$

of states $|S_i\rangle$ with weights c_i

$$|S\rangle = \sum_i c_i |S_i\rangle, \quad \sum_i |c_i|^2 = 1. \tag{10.2}$$

Then, because of the linearity of quantum theory, the final state $\mathcal{S} + \mathcal{A}$ is

$$|\Psi\rangle = \sum_i c_i |S_i\rangle |A_i\rangle, \tag{10.3}$$

that is, an entangled state. Reading the apparatus does give the result $|S_i\rangle$ with probability $|c_i|^2$, in conformity with Born's rule, but in the state (10.3), *no measurement has been performed*: we still have a quantum superposition, which contains only a potentiality of results, but not the results themselves. Moreover, if the apparatus is macroscopic, equation (10.3) describes a superposition of macroscopically distinguishable states, called a *Schrödinger cat*. This terminology originates in a situation imagined by Schrödinger in 1935, with the purpose of underlining its absurdity. A cat is enclosed in a box which contains also an atom with a radioactive nucleus that decays with a half life of one hour. If the nucleus decays, a devilish device kills the unfortunate cat: one of the decay products triggers a detector, itself linked to a hammer which breaks a flask containing a lethal poisonous gas which leads instantaneously to the cat death. The state of the cat is entangled with that of the nucleus as in (10.3), and after one hour, the quantum state is a superposition with equal weights of a first quantum state corresponding to the non-decayed atom $|ATOM\rangle$ and to the live cat $|CAT\rangle$, and of a second one corresponding to the decayed atom $|atom\rangle$ and to the dead cat $|cat\rangle$

$$|\Psi\rangle = \frac{1}{\sqrt{2}}(|ATOM\rangle |CAT\rangle + |atom\rangle |cat\rangle).$$

von Neumann's theory has the unavoidable consequence that we must treat measuring apparatuses as quantum objects, which may display superposition of states seemingly absurd and never observed. To sum up, von Neumann's theory faces two problems: (i) equation (10.3) does not describe a measurement which is actually completed, but only a premeasurement, and (ii) it inevitably introduces superpositions of macroscopically distinguishable states.

Box 10.5. Minimalist interpretation and Schrödinger's cat.

For a proponent of the minimalist interpretation, there is no problem at all except that he might get into trouble with the Society for the Prevention of Cruelty to Animals (SPCA). The experiment is done on a large number (ensemble) of cats, and after one hour, half of the cats are alive and the other half is dead. Quantum theory correctly predicts the probability of each of the two possibilities, and this is all we may require from it. It is meaningless to ask for the fate of cat number 8. The paradox arises only when we wish to apply quantum theory to individual events.

It at this point that von Neumann introduces the observer. It is when he or she reads the apparatus that the observer registers the result in his or her brain in an indelible way, and this process corresponds to the irreversible recording of the measurement. This aspect of von Neumann's theory is not taken seriously any longer today, one good reason being that detector clicks are registered automatically with computer control, and human observation takes place only much later when reading computer screens and when all results have been registered on hard drives a long time ago. In a similar way, the (possible) death of Schrödinger's cat does not take place when the observer opens the box, but at the time when the detector registers in an irreversible way the particle emitted in the radioactive decay and triggers the devilish device. However, the first part of von Neumann's theory, that which leads to (10.3), is still the basis of modern theories of quantum measurement today.

10.3 The measuring apparatus is macroscopic

Wave function collapse is a rather unsatisfactory characteristic of von Neumann's theory, or of the orthodox interpretation, which we would like to get rid of while keeping only the deterministic part, the Schrödinger equation. von Neumann's theory does not give any specification on the measuring apparatus, except that it is governed by quantum laws. However, even if the measurement process begins with a microscopic interaction between the system S and the apparatus A, the macroscopic character of the latter could well play a major role. In the famous text by Landau and Lifschitz (Landau was intellectually very close to Bohr), which begins with a presentation of the arguments leading to (10.3), the problem is "solved" by decree by imposing the uniqueness of the apparatus state,

"The classical character of the apparatus is expressed in the property that, at each instant of time, we can assert positively that it is in a known state".

which is essentially Bohr's point of view, as the classical character of the apparatus plays a fundamental role. As a consequence, the measurement result, which is in one-to-one correspondence with the state of the apparatus, is *unique*, so that there is uniqueness of the outcome. In equation (10.3), the apparatus can only be in a unique state $|A_i\rangle$, and not in a superposition of such states.

Models grounded in decoherence, which try to bypass this arbitrary decree, are quite popular today. We recall that *decoherence is a consequence of the deterministic equation of evolution, and its framework is strictly contained within quantum theory*: see Chapter 9. A macroscopic measuring apparatus is endowed with an enormous number of degrees of freedom, much larger than in the case of the fullerene molecules in Chapter 9, and these degrees of freedom are only asking for interaction with the environment: molecules in the air, vibrations, thermal motion etc. The description which follows is schematic, and there are much more sophisticated models which, however, all rely on the same strategy. von Neumann's approach is generalized by considering the apparatus as being embedded in an environment $\mathcal{E}$, treated as a quantum system and whose quantum states are denoted by $|E_i\rangle$. The composite $\mathcal{S}+\mathcal{A}+\mathcal{E}$ system is initially in a state $|\Phi\rangle$

$$|\Phi\rangle = \left(\sum_i c_i |S_i\rangle \right) |A_0\rangle |E_0\rangle,$$

where $|A_0\rangle$ is as previously the initial "neutral" state of the apparatus and $|E_0\rangle$ the initial state of the environment. For simplicity, but this is not essential, we assume that the $\mathcal{S}\mathcal{A}$ interaction takes place first

$$|\Phi\rangle \rightarrow |\Phi'\rangle = \left(\sum_i c_i |S_i\rangle |A_i\rangle \right) |E_0\rangle.$$

The model is built in such a way that the $\mathcal{A}\mathcal{E}$ interaction has an effect similar to that of the $\mathcal{S}\mathcal{A}$ one in the preceding section

$$|A_i\rangle |E_0\rangle \rightarrow |A_i\rangle |E_i\rangle.$$

Once all the interactions are completed, the global state $|\Psi\rangle$ is

$$|\Psi\rangle = \sum_i c_i |S_i\rangle |A_i\rangle |E_i\rangle. \tag{10.4}$$

We recover once more the situation of Chapter 9: if the states $|E_i\rangle$ of the environment are distinguishable, that is, orthogonal, $\langle E_i|E_j\rangle = \delta_{ij}$, then we shall observe the state $|A_i\rangle$ with a probability $|c_i|^2$. As in the case of the interferometer with vertically polarized photons in one arm and horizontally polarized ones in the other, the coherence explicit in the expression (10.3) $\sum_i c_i |S_i\rangle |A_i\rangle$ has disappeared, and the same thing has happened to the catastrophic superpositions of macroscopically distinguishable states of the apparatus, or Schrödinger cats. The coherences contained in (10.4) are now distributed over an enormous number of degrees of freedom of the environment, which makes the process irreversible in practice, even though the global state vector $|\Psi\rangle$ is still, in principle, governed by a reversible evolution. The quantum correlations of the $\mathcal{SA}$ system have been transferred to the environment, and an *observation* limited to $\mathcal{A}$ will reveal the absence of interference. Moreover, only the states of the apparatus which are stable under their interaction with the environment, that is, which do not entangle with it, may appear in the superposition (10.4). These states span a basis privileged by decoherence, because they are resilient to it. They are termed classical, or *pointer states*, as they may represent a pointer on a dial. In many cases, position-localized states are pointer states, and in quantum optics, pointer states are the so-called coherent states. May we conclude that there are no longer Schrödinger cats, that the apparatus is in a well-determined state and that the quantum measurement problem is solved? Unfortunately, this is not the case, but the reader must be warned that the preceding negative statement is still hotly debated and opposite statements may be found in the literature. In order to understand our caveat, let us return to the example of the interferometer with polarized photons: as long as the photon has not been registered by a detector, that is, observed, we cannot say that the photon followed one of the two paths, because entanglement with polarization is not enough to determine the path. This determination is effective *only when the photon has been registered.* When we go from a microscopic quantum object to a macroscopic one (the apparatus), *the rules of quantum physics do not change*: this point was strongly emphasized by the Nobel prize winner Anthony Leggett. Entanglement with the environment is not able, by itself, to force the apparatus into a unique state. The fact that the environment is of enormous complexity does not matter: as long as the $\mathcal{S} + \mathcal{A} + \mathcal{E}$ evolution is deterministic, we cannot get a definite result for the measurement. Decoherence does not account for the uniqueness of outcomes.

The above argument is important, but controversial. It is then useful to give a second version, which is physically equivalent, but which differs in detail. It is grounded on the distinction between two different kinds of probabilistic ensembles. In the example of the Mach–Zehnder interferometer of Section 9.1, with polarization insensitive detectors, we know, for any specific photon of the ensemble *after detectors have been triggered*, that this photon followed *definitely* one of the paths, even though we cannot tell which one. In this case, we are dealing with a classical probabilistic mixture, where each individual in the ensemble is in a well-defined state, even though this state remains unknown to us. Our only information is the *probability* that an individual in the ensemble be in such or such state. On the other hand, *as long as the photon has not been detected*, we can only assert that a polarization sensitive detector would give us the which-path information, each of these paths having a 50% probability, but this path does not exist before a detector has been triggered. There were two possibilities before detection, the red path and the blue path (Figure 9.1), but neither has been realized. This second kind of probabilistic ensemble is of the quantum kind. These two ensembles have been termed by Bernard d'Espagnat "proper probabilistic ensembles" (classical) and "improper probabilistic ensembles" (quantum). The probabilistic ensemble of measurements performed by $\mathcal{A}$ and summarized in (10.3), and when we do not measure the environment (in physicist speak: when we take the partial trace over the degrees of freedom of the environment), is a probabilistic ensemble of the quantum kind. As a consequence, no possibility for the pointer to show a definite direction on the dial has been realized.

10.4 Non-standard interpretations

Wave function collapse is, as we have seen, the most questionable part of the orthodox formalism. Several proposals (or interpretations) have been introduced in order to avoid it. Let us say right away that all these proposals meet with such difficulties that none of them takes a decisive edge over the standard interpretation. We shall limit ourselves to two of them.

1. The hidden variables theory first proposed by de Broglie and improved by Bohm.
2. Everett's interpretation and its many-worlds variant.

The *de Broglie–Bohm theory* (Appendix A.10.1) introduces two coupled elements, a particle and a wave function, which define completely the state of a microscopic system. Note that we need the coexistence of *two* elements in order to describe a *single* particle. A particle is always endowed with a position and a velocity (or a momentum) which are well defined. The particle position plays a privileged role: its evolution is governed by a deterministic equation similar to Newton's second law. However, the potential energy from which the force is derived is not reduced to that of classical mechanics. A quantum potential, proportional to Planck's constant squared, must be added to the standard expression, and this potential vanishes in the classical limit $h = 0$. This quantum potential is determined by the wave function $\psi(\vec{r}, t)$ which depends on the particle position $\vec{r}$ and on the time t and obeys the standard Schrödinger equation (11.43). It influences the particle trajectory, but there is no feedback of the particle on the wave, so that there is an asymmetry between the two elements of the description. The particle is an element of a probabilistic ensemble and the square of its wave function is *assumed*, in a somewhat *ad hoc* way, to give the probability of finding it at some point in space. The momentum probability distribution is computed as in the standard theory, so that the position and momentum distributions measured on an ensemble of particles obey Heisenberg's inequalities (4.3). However, this probability is of the ignorance kind, due to our imperfect knowledge of the particle position, and not an intrinsic probability of the quantum kind. The particle position is a hidden variable, which can be experimentally measured, but which we cannot control or manipulate, while it is possible to act on the wave function by modifying the external potential. The theory is *realistic*: position and wave function exist independently of observation; they are part of an external reality.

The quantum potential displays an essential characteristic: it is non-local, in the sense that a modification of the wave function in a given point of space affects *instantaneously* a distant point. It is illuminating to look at a Young slit experiment (Figure 1.9) in this theory. Each particle of the ensemble follows a well defined path and goes through one of the slits. This seems to contradict the discussion of Section 1.4, according to which interferences are destroyed when the path is determined: we may then divide the particles of an ensemble in two classes, those who went through slit S_1 and for which we could have as well closed slit S_2, and vice-versa. But this reasoning is not valid in de Broglie–Bohm's theory, because the quantum potential which guides the particle is not the same depending on whether only one or the two slits are open. The particle going through slit S_1 is sensitive to the fact that slit S_2 is open or closed, because the wave $\psi(\vec{r}, t)$ is not the same in both cases. Because of the quantum potential, a particle may exhibit a curved trajectory in free space, even in the absence of forces, a phenomenon at the origin of the interference pattern.

The results of EPR-like experiments with non-relativistic particles are perfectly explained by this theory, at the price of rather cumbersome calculations. The calculation is done for spin 1/2 particles: spin, or proper angular momentum, is a property similar to polarization, and spin 1/2 particles have two spin states, for example up and down, the analogue of vertical and horizontal polarization states of photons. The spin measurements are performed thanks to magnets able to deflect the particles according to their spin orientation, much as PBSs deflect photons according to their polarization. Non-locality is essential in order to bring theory into agreement with experiment. When Alice's spin is deflected by a magnet, the deviation exerts instantaneously an influence on Bob's spin, and forces it to follow a trajectory in agreement with that predicted by standard quantum mechanics. As we have mentioned, the de Broglie–Bohm theory is a realistic theory: particles are endowed with properties in the absence of observation. Wave function collapse follows from the theory and does not have to be imposed in an *ad hoc* fashion: a measurement only reveals the pre-existing position of the particle. The theory obeys separability: each of the spin 1/2 particles in an EPR-like experiments keeps its own identity, but it is non-local. It reproduces the results of standard quantum mechanics, but at the price of a cumbersome formalism. It does not make any new prediction and, mostly, it does not seem to have a relativistic generalization, although this last statement is disputed. In spite of that, it is still favored by an active minority of physicists.

Everett's interpretation was first proposed by Everett and then clarified by Wheeler, hence the often encountered terminology: Everett–Wheeler interpretation. It relies on an entirely different idea in order to avoid wave function collapse. Let us assume that the state vector of the system S to be measured is in the superposition (10.2): $|S\rangle = \sum_i c_i |S_i\rangle$. An observer $\mathcal{O}$ — who is not necessarily a human observer, but may be an automatic recording device — performs a measurement which collapses the wave function: $|S\rangle \rightarrow |S_i\rangle$. How is this process perceived by an observer $\mathcal{P}$ who is linked neither to S, nor to $\mathcal{O}$? This observer ascribes to $\mathcal{O}$ before the measurement an initial state vector $|\mathcal{O}_{\text{init}}\rangle$, and he or she describes the measurement thanks to equation

$$\sum_i c_i |S_i\rangle |\mathcal{O}_{\text{init}}\rangle \rightarrow \sum_i c_i |S_i\rangle |\mathcal{O}_i\rangle. \tag{10.5}$$

We notice the formal analogy with (10.3): the right hand side of (10.5) defines an entangled state, and $|\mathcal{O}_i\rangle$ is the state of the observer who obtained the result $|S_i\rangle$. Everett interprets (10.5) as the property that $\mathcal{O}$ is in state $|\mathcal{O}_i\rangle$ *relative* to $|S_i\rangle$: this is the reason why Everett's interpretation is sometimes called the relative state interpretation. Wave function collapse is valid but only at the subjective level of each of the observers featured in the super-position (10.5): each term in the superposition describes an observer who perceived a definite result, but the result is different for another observer. Wave function collapse appears as subjectively valid for $\mathcal{O}$, but no collapse occurs for observer $\mathcal{P}$.

However, as the perceptions by observer $\mathcal{O}$ represented in the different terms of (10.5) are contradictory, how can they be simultaneously realized? As Everett never answered this question unambiguously, many versions of his interpretation have been proposed. The most popular one is the many-worlds interpretation of Graham and de Witt. According to these authors, at the very instant when a measurement is performed, the Universe splits into multiple parallel universes, and each term in the decomposition (10.5) is effectively realized in one of these universes. Each of the observers lives in a specific world, and he or she cannot access the parallel universes. In addition to some technical difficulties, for example problems with relativistic invari-ance and the definition of a privileged basis which is implicit in (10.5), this interpretation is at odds with common sense: is it not paradoxical to require

10^{100} copies, at least, of our Universe in order to solve the problem of wave function collapse, and is the solution brought by the many-worlds interpretation not wilder than the logical contradiction which is found today in this collapse? However, the boldness of this interpretation cannot be taken as a refutation, and, according to a recent poll, it attracts a majority of cosmologists who find the existence of parallel universes appealing.

10.5 Conclusion

We are now at the end of this guided tour of quantum physics. I hope to have convinced the reader who was able to follow me up to this point that quantum theory is really universal and has proven to be extraordinarily efficient. The modification of the quantum physics landscape in the past thirty years is spectacular: in 1980, the experimental status of Bell's inequalities was shaky, neither cold atoms nor atomic Bose–Einstein condensates existed, quantum cryptography was yet to be discovered. Neither laser diodes nor night vision devices in the infrared were available commercially, and entirely artificial materials for electronics and opto-electronics were still in their infancy. In the years to come, we may expect spectacular experimental advances which will allow us to better understand the quantum-to-classical frontier, and (maybe) to better figure out the meaning of quantum theory. There does not yet exist a consensus on the meaning of concepts introduced by the theory, but the end of the last century witnessed a challenge of the until-then dominant Copenhagen interpretation. Two decoherence theorists, Eric Joos and Maximillian Schlossauer, may then write

> Joos: "Classical properties are then not an *a priori* attribute of objects, but only come into being through the irreversible interactions with the environment... Traditional, but ill-defined concepts such as dualism, Heisenberg uncertainty relations or complementarity appear obsolete from this point of view."

> Schlossauer: "To lift the frog from the North (that is, Copenhagen!) we should see the wave function as more than a tool for calculating probabilities."

However many physicists, and important ones, still support Bohr's interpretation, for example, the late John Wheeler, David Mermin or Anthony Leggett.

Wheeler: "The Copenhagen interpretation remains the best interpretation of quantum mechanics."

Mermin: "Admittedly, you can't entirely eliminate the discomfort that gives rise to "quantum non-locality" and the "measurement problem" by acknowledging that quantum sates are not real properties of the system they describe. But the recognition that quantum states are not real properties of a system forces one to formulate the source of the discomfort in more nuanced, less sensational terms."

Leggett: "QM is the complete truth about the physical world (in the sense that it will always give reliable experimental predictions) but it describes no external reality."

These are only a few quotations, but they suffice to show that the status of quantum theory is still hotly debated! Quantum mechanics is probably far from being completed, contrary to the case of classical mechanics which was finalized one century ago with its relativistic version. It promises to stay alive and rich in new advances for a long time.

10.6 Further reading

There exist dozens of books devoted to the interpretation of quantum mechanics and the measurement problem; among them, I would recommend d'Espagnat [2003] and Laloë [2012]. Non-destructive quantum measurements were performed, for example, by Guerlin et al. [2007] and Gleyzes et al. [2007]. The textbooks quoted in the text are those by Peres [1993], Ballentine [1998] and Landau and Lifschitz [1977]. Howard [2004] gives an interesting overview of the Copenhagen interpretation, while Leggett's article [2005] is a short, but incisive, summary of the measurement problem. The experiment on wave-particle duality was performed by Kaiser et al. [2012]. The standard reference on de Broglie–Bohm theory is Holland [1993].

11

Appendices

A.1.1 Units

Physicists, like other scientists, need to manipulate very large or very small numbers. This can become quite cumbersome if we do not use appropriate units, obtained thanks to prefixes and their abbreviations. Let us first examine large numbers.

- $1000 = 10^3 = $ kilo (k)
- $1\,000\,000 = 10^6 = $ mega (M)

For example, 1 kilometer $= 10^3$ m $= 1$ km, 1 mega electron-volt $= 10^6$ eV $= 1$ MeV. Further large units are.

- $10^9 = $ giga (G)
- $10^{12} = $ tera (T)
- $10^{15} = $ peta (P)

For example, 1 terawatt $= 1$ TW $= 10^{12}$ W.

Now for small numbers.

- $0.001 = 10^{-3} = $ milli (m)
- $0.000\,001 = 10^{-6} = $ micro (μ)

Table 11.1.   Orders of magnitude in meters for some typical distances.

Known Universe	Galaxy radius	Sun–Earth distance	Earth radius	Man	Insect
$\sim 10^{27}$	$\simeq 5 \times 10^{20}$	1.5×10^{11}	6.4×10^6	$\simeq 1.7$	0.01 to 0.001
E. coli bacterium	VIH virus	fullerene C_{60}	atom	lead nucleus	proton
$\simeq 2 \times 10^{-6}$	1.1×10^{-7}	0.7×10^{-9}	$\simeq 10^{-10}$	7×10^{-15}	0.8×10^{-15}

For example, 1 milliwatt $= 1\,\text{mW} = 10^{-3}\,\text{W}$, 1 micrometer $= 1\,\mu\text{m} = 10^{-6}$ m. Further small units are.

- $10^{-9} = $ nano (n)
- $10^{-12} = $ pico (p)
- $10^{-15} = $ femto (f)

For example, 1 nanometer $= 1\,\text{nm} = 10^{-9}$ m, 1 femtosecond $= 1\,\text{fs} = 10^{-15}$ s.

It is instructive to use the exponential notation for Table 11.1, which gives typical order of magnitudes for lengths encountered in the Universe, from the dimensions of the known Universe down to elementary particles. It is useful to remember that the size of an atom is about 0.1 nm, or 10^{-10} m, and that of an atomic nucleus is a few fm (10^{-15} m).

A.1.2 Electromagnetic waves and protons

Electromagnetic waves span a large domain, which goes from radio waves to the waves of γ-radioactivity (Figure 11.1). Actually, as in the latter case the particle aspect is particularly obvious, one usually speaks of γ-photons, while the photon concept is rarely used for radio waves, although there is strictly no reason of principle behind this common practice. Let us recall that, despite their apparent differences, all these waves are identical: viewed by an observer moving at very high speed with respect to the Earth, a radio wave could appear as a γ-ray, as a consequence of the Doppler effect (Chapter 5).

Light waves propagate in a vacuum at a speed $c = 2\,999\,792\,558$ m/s $\simeq 3 \times 10^8$ m/s. It would take 0.13 s for a light ray to make a round trip around the Earth at the equator. In a transparent medium like water or glass, the

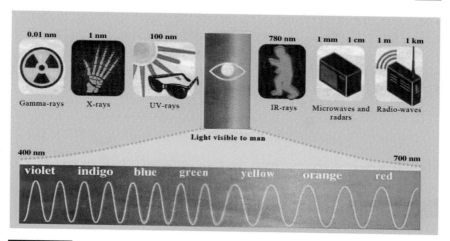

Figure 11.1. Spectrum of the electromagnetic radiation. The shortest wavelengths are those of γ-rays, and then the X-rays and the UV rays. We then find the visible domain, which extends from violet to red. Increasing further the wavelength, we first enter the infrared domain, then that of microwaves, and finally that of radio waves. The wavelength is given in nanometers (nm), $1 \, \text{nm} = 10^{-9}$ m.

speed is reduced by n, where n is the optical index, $n \simeq 1.3$ for water, $n \simeq 1.5$ for glass. It would take $0.13 \, \text{s} \times n$ for a light ray to make a round trip around the Earth at the equator in an optical fiber with index $n \simeq 1.48$. Light waves are a particular case of electromagnetic waves, whose spectrum is represented in Figure 11.1. One important characteristic of an electromagnetic wave is its period T. It is more usual to characterize an electromagnetic wave by its frequency $\nu = 1/T$, which is measured in Hertz (Hz). A wave is repeated periodically in space and time, and the spatial periodicity is the wavelength $\lambda = cT = c/\nu$: it is the distance between two successive crests (maxima) or two successive troughs of a propagating wave. The wavelengths of light lie between $\lambda \simeq 0.40 \, \mu\text{m}$ (violet) and $\lambda \simeq 0.70 \, \mu\text{m}$ (red). As in Figure 11.1, wavelengths are often measured in nm. For $\lambda < 0.40 \, \mu\text{m}$ we enter the ultraviolet domain, and for $\lambda > 0.70 \, \mu\text{m}$, the infrared one. Note also the importance of the $1.31 \, \mu\text{m}$ and $1.55 \, \mu\text{m}$ wavelengths in the infrared, which are almost universally used in optical fibers.

The energy of a photon corresponding to a wavelength λ is given by

$$E = \frac{hc}{\lambda} \tag{11.1}$$

where h is Planck's constant. The shorter the wavelength, and the larger the energy: a γ-photon is more energetic that a X-ray photon, which is itself more energetic than an ultraviolet photon etc. Energy can also be expressed as a function of the frequency ν, which is given as a function of the wavelength by $\nu = c/\lambda$. The energy of a photon is then given by the Planck–Einstein relation

$$E = h\nu = \hbar\omega \qquad (11.2)$$

where $\hbar = h/2\pi$ and $\omega = 2\pi\nu$ is the angular frequency (Figure 11.2).

A.1.3 Trigonometric functions and complex numbers

In order to define trigonometric functions, we draw a circle of unit radius and denote by OM the radius joining the center O to some point M on the circumference (Figure 11.3). The angle between OM and the horizontal axis OH is θ. Let $\overline{OA}$ and $\overline{OB}$ be the algebraic projections on the horizontal and vertical axis, respectively. By definition

$$\overline{OA} = \cos\theta, \qquad \overline{OB} = \sin\theta.$$

In the case of Figure 11.3, $\cos\theta$ and $\sin\theta$ are both positive, but they may have in general a positive or a negative value. Pythagoras's theorem entails

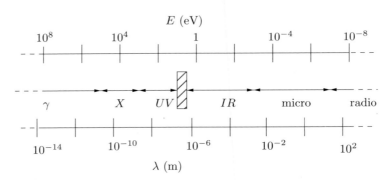

Figure 11.2. Wavelengths of electromagnetic radiation and energy of the corresponding photons. The hatched box represent the visible domain. Frontiers between the different kinds of radiation, for example, the frontier between X-rays and γ-rays, are not sharply defined. A photon with energy $E = 1\,\text{eV}$ in the near infrared has a wavelength $\lambda = 1.24 \times 10^{-6}\,\text{m}$, a frequency $\nu = 2.42 \times 10^{14}\,\text{Hz}$ and an angular frequency $\omega = 1.52 \times 10^{15}\,\text{rad}\cdot\text{s}^{-1}$.

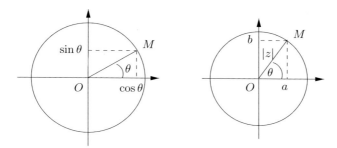

Figure 11.3. (a) A circle with unit radius defines the sine and cosine functions. (b) Geometrical representation of a complex number. The circle radius is $|z|$, the real part of z is $a = |z| \cos \theta$ and its imaginary part $b = |z| \sin \theta$.

the relation

$$\cos^2 \theta + \sin^2 \theta = 1. \qquad (11.3)$$

A *complex number* z is of the from

$$z = a + ib$$

where a and b are two real numbers, positive or negative, and $i^2 = -1$; a is the *real part* of z and b its *imaginary part*. Calculations rules are those of usual algebra, for example if $z_1 = a_1 + ib_1$ and $z_2 = a_2 + ib_2$, the product $z_1 z_2$ is given by

$$z_1 z_2 = (a_1 a_2 - b_1 b_2) + i(a_1 b_2 + b_1 a_2).$$

The absolute value of z, also called its *modulus*, is

$$|z| = \sqrt{a^2 + b^2}. \qquad (11.4)$$

There exists another useful way of writing z. The complex conjugate of z, noted z^*, is defined by $z^* = a - ib$, (mathematicians use the notation $\bar{z}$, instead of z^*), and

$$|z|^2 = a^2 + b^2 = (a + ib)(a - ib) = zz^*. \qquad (11.5)$$

We may now give a convenient representation of a complex number: the horizontal axis corresponds to the real part, the vertical one to the imaginary part, and the vector $\overrightarrow{OM}$ has components a and b along these axes. The length of $\overrightarrow{OM}$ is nothing other than the modulus $|z|$. If θ is the angle between the vector $\overrightarrow{OM}$ and the horizontal axis, we have, by the very definition of trigonometric functions

$$a = |z| \cos \theta, \quad b = |z| \sin \theta. \tag{11.6}$$

Moreover, there exists a remarkable relation between trigonometric functions and the exponential function (Moivre's formula)

$$e^{i\theta} = \cos \theta + i \sin \theta, \tag{11.7}$$

or, conversely

$$\cos \theta = \frac{1}{2}(e^{i\theta} + e^{-i\theta}), \quad \sin \theta = \frac{1}{2i}(e^{i\theta} - e^{-i\theta}).$$

We may thus rewrite z in the form

$$z = |z|(\cos \theta + i \sin \theta) = |z| e^{i\theta}. \tag{11.8}$$

The angle θ is the *phase* of the complex number, and equation (11.8) gives a modulus/phase representation of a complex number z. Let us give as an example the calculation of the interference pattern for the Mach–Zehnder interferometer (Appendix A.1.4) where we need $|\exp(i\theta) + 1|^2$. Using (11.5) we obtain

$$|e^{i\theta} + 1|^2 = [e^{i\theta} + 1][e^{-i\theta} + 1]$$
$$= 1 + e^{i\theta} + e^{-i\theta} + 1 = 2(1 + \cos \theta).$$

Formula (11.7) allows us to derive important trigonometric relations, for example

$$e^{2i\theta} = \cos 2\theta + i \sin 2\theta = (\cos \theta + i \sin \theta)^2$$
$$= (\cos^2 \theta - \sin^2 \theta) + 2i \sin \theta \cos \theta,$$

whence, by identification of real and imaginary parts

$$\cos 2\theta = \cos^2 \theta - \sin^2 \theta = 2 \cos^2 \theta - 1, \quad \sin 2\theta = 2 \sin \theta \cos \theta.$$

Finally, we derive (1.5), by computing the modulus of $z = z_1 + z_2$

$$|z|^2 = |z_1|^2 + |z_2|^2 + 2|z_1 z_2| \cos \delta$$

where δ is the angle between the vectors $\overrightarrow{OM_1}$ and $\overrightarrow{OM_2}$ corresponding to z_1 and z_2, respectively.

A.1.4 Theory of the Mach–Zehnder interferometer

In this appendix, we give the theory of the Mach–Zehnder interferometer by using complex numbers. The beamsplitter BS_2 combines wave amplitude coming from the blue path and the red path in Figure 1.6. In the case of the a_X amplitude corresponding to the horizontal path after BS_2, the amplitudes coming from the blue path and that coming from the red one have both experienced a reflection and a transmission: reflection by BS_1 and transmission by BS_2 for the blue path, and vice-versa for the red one. For simplicity, we assume that the two arms of the interferometer have exactly the same length. Then the only difference comes from the phase shift $\delta = 2\pi \ell / \lambda$ owing to the additional distance on the blue path. In the absence of this phase shift, both amplitudes would be equal to $a/2$, if we take into account the division by $1/\sqrt{2}$ at each beamsplitter. With this phase shift included, the amplitude coming from the blue path is $a \exp(i\delta)/2$, and we get for a_X

$$a_X = \frac{a}{2} \left(e^{i\delta} + 1 \right), \tag{11.9}$$

from which we deduce the intensity I_X

$$I_X = |a_X|^2 = \frac{|a|^2}{4} |e^{i\delta} + 1|^2 = \frac{I}{2} (1 + \cos \delta), \tag{11.10}$$

where we have used (Appendix A.1.3)

$$|e^{i\delta} + 1|^2 = (e^{i\delta} + 1)(e^{-i\delta} + 1) = 2(1 + \cos \delta).$$

It is interesting to comment on the origin of the minus sign in $I_Y = I(1 - \cos \delta)/2$. Actually, in the calculation of a_Y, we observe that the amplitude coming from the blue path experienced two reflections, and that coming from the red path experienced two transmissions. One can show in optics that each reflection multiplies the amplitude by a factor i, and there is thus a difference by a factor $i^2 = -1$ between the contributions of the two paths, so that

$$a_Y = \frac{a}{2}(-e^{i\delta} + 1),$$

hence the value of I_Y.

A.2.1 RSA encryption

Bob chooses two primes p and q, $N = pq$, and a number c having no common divisor with the product $(p - 1)(q - 1)$. He calculates d, the inverse of c for mod $(p - 1)(q - 1)$ multiplication:

$$cd \equiv 1 \bmod (p - 1)(q - 1).$$

By a non-secure channel he sends Alice the numbers N and c (but not p and q separately!). Alice wants to send Bob a message, which must be represented by a number $a < N$ (if the message is too long, Alice can split it into several sub-messages). She then calculates (Figure 11.4)

$$b \equiv a^c \bmod N$$

and sends b to Bob, always by a non-secure channel, because a spy who knows only N, c, and b cannot infer the original message a. When Bob

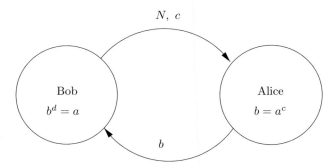

Figure 11.4.   RSA encryption scheme. Bob chooses $N = pq$ and c. Alice encrypts her message a using $b = a^c \bmod N$ and Bob decrypts it using $b^d = a \bmod N$.

receives the message he calculates

$$b^d \bmod N = a.$$

The fact that the result is precisely a, that is, the original message of Alice, is a result of number theory. To summarize, the numbers N, c, and b are sent publicly, by a non-secure channel.

Example.

$$p = 3, \quad q = 7, \quad N = 21, \quad (p-1)(q-1) = 12.$$

The number $c = 5$ has no common factor with 12, and its inverse with respect to mod 12 multiplication is $d = 5$ because $5 \times 5 = 24 + 1$. Alice chooses $a = 4$ for her message. She calculates

$$4^5 = 1024 = 21 \times 48 + 16, \quad 4^5 = 16 \bmod 21.$$

Alice then sends Bob the message 16. Bob calculates

$$b^5 = 16^5 = 49\,932 \times 21 + 4, \quad 16^5 = 4 \bmod 21,$$

thus recovering the original message $a = 4$. The above calculation of $16^5 \bmod 21$ has not been done very cleverly. There are much faster methods allowing one to manipulate only numbers which are not large compared to N.

A.2.2 Mathematical description of polarization

The *polarization of light* was demonstrated for the first time by the Chevalier Malus in 1809. He observed the light of the setting sun reflected by the glass of a window in the Luxembourg Palace in Paris through a crystal of Iceland spar and discovered that when the crystal was rotated, one of the two images of the sun disappeared. Iceland spar is a birefringent crystal which decomposes a light ray into two rays polarized in perpendicular directions. As the ray reflected from the glass is (partially) polarized, when the crystal is suitably oriented, one observes the disappearance (or strong attenuation) of one of the two rays.

The scheme in Figure 2.1 suggests a simple geometrical interpretation of polarization. Let us begin with the wave description of light. As

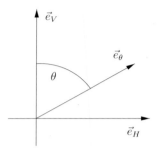

Figure 11.5. The polarization vector of light.

polarization defines a direction in a plane perpendicular to the beam, we may interpret polarization as a vector $\vec{e}_\theta$ of unit length ($\vec{e}_\theta^2 = 1$) in this plane. If $\vec{e}_V$ and $\vec{e}_H$ denote two perpendicular unit vectors in this plane, respectively vertical and horizontal, any polarization may be written as (Figure 11.5)

$$\vec{e}_\theta = \vec{e}_V \cos\theta + \vec{e}_H \sin\theta. \qquad (11.11)$$

The quantities $\cos\theta$ and $\sin\theta$ are *wave amplitudes*: if I_0 is the intensity entering the polaroid, the intensity exiting a polaroid with a vertical axis is $I_V = I_0 \cos^2\theta$ and $I_H = I_0 \sin^2\theta$ if the polaroid has a horizontal axis. The proportionality to $\cos^2\theta$ is the *Malus law*. A particularly interesting case is that where $\theta = 45°$, with $\cos\theta = \sin\theta = 1/\sqrt{2}$ and $I_V = I_H = I_0/2$.

We now tackle the case of photon polarization. Let us assume that we have a source of polarized single photons, but we do not know their polarization. We send these photons into a polaroid of vertical axis. Some of the photons are stopped, but those which go through the polaroid will have their polarization along $\vec{e}_V$. If the polaroid axis makes an angle θ with the vertical, the photons which go through this polaroid will have their polarization aligned along $\vec{e}_\theta$ (11.11). In quantum physics, we say that the polaroid has *prepared* the photons in a (quantum) state of polarization $\vec{e}_\theta$: this is the *preparation stage* of an experiment in quantum physics. If these photons are sent into a polaroid with vertical (horizontal) axis, they will be transmitted with a probability $\cos^2\theta$ ($\sin^2\theta$): we must recover the wave result, that is, the Malus law, in the limit of a large number of photons. The quantity $\cos\theta$ (resp. $\sin\theta$) is the *probability amplitude* of finding photons with polarization $\vec{e}_\theta$ in a state of vertical (horizontal) polarization. According to the rules stated in Chapter 1, probabilities are given by amplitudes squared. The following remark is going to play an important role: the probability amplitudes $\cos\theta$ and $\sin\theta$ are the scalar products of $\vec{e}_\theta$ with $\vec{e}_V$ and $\vec{e}_H$

$$\cos\theta = \vec{e}_V \cdot \vec{e}_\theta, \quad \sin\theta = \vec{e}_H \cdot \vec{e}_\theta. \tag{11.12}$$

Let us recall that the scalar product $\vec{a} \cdot \vec{b}$ of two vectors $\vec{a}$ and $\vec{b}$ is by definition

$$\vec{a} \cdot \vec{b} = \|a\| \|b\| \cos\phi,$$

where $\|a\|$ and $\|b\|$ are the lengths of the two vectors and ϕ the angle between them.

Dirac introduced a compact notation for quantum states, which we may apply to polarization states $\vec{e}_V, \vec{e}_H$ and $\vec{e}_\theta$, which, in this notation, are written $|V\rangle$, $|H\rangle$ and $|\theta\rangle$. Equation (11.11) becomes

$$|\theta\rangle = \cos\theta |V\rangle + \sin\theta |H\rangle, \tag{11.13}$$

while the scalar product $\vec{e}_V \cdot \vec{e}$ is denoted $\langle V|\theta\rangle$. As $|V\rangle$, $|H\rangle$ and $|\theta\rangle$ describe quantum states, they are called *polarization state vectors* of photons polarized along V, H or θ. The vector states of photons polarized at $\pm 45°$ used in the BB84 protocol implemented with polarized photons are then

$$|D\rangle = \frac{1}{\sqrt{2}}(|V\rangle + |H\rangle),$$
$$|A\rangle = \frac{1}{\sqrt{2}}(|V\rangle - |H\rangle). \tag{11.14}$$

To conclude, we mention a slight complication: we saw in Chapter 1 that probability amplitudes are complex numbers. The most general polarization state vector

$$|\varphi\rangle = \cos\theta |V\rangle + e^{i\delta} \sin\theta |H\rangle \tag{11.15}$$

depends, in addition to $\cos\theta$ and $\sin\theta$, on the complex number $\exp(i\delta)$. In the general case described by (11.15), a photon polarization is elliptic which means that the tip of the polarization vector describes an ellipse in the plane perpendicular to the direction of propagation. Note that a particular, and important, case of elliptic polarization is that of circular polarization, right (R) or left (L)

$$|R\rangle = \frac{1}{\sqrt{2}}(|V\rangle + \mathrm{i}|H\rangle),$$

$$|L\rangle = \frac{1}{\sqrt{2}}(|V\rangle - \mathrm{i}|H\rangle).$$

(11.16)

The tip of the polarization vector traces a circle in the plane perpendicular to the direction of propagation. If we look at the photon coming toward us, we see this vector rotating clockwise in the case of left-handed circular polarization, and counterclockwise for right-handed polarization.

A.2.3 Attenuated light

Let us return to the example of Figure 1.4, where a light wave is incident on a balanced beamsplitter, and let us decrease the light intensity such that the time interval between two successive photons is larger than the dead time between two detections. We then register isolated clicks of the detectors. Let $\langle n\rangle$ be the average number of photons incident on the beamsplitter during some well-defined time interval, for example one micro-second, and suppose $\langle n\rangle = 0.1$. It can be shown that in the case of light emitted by a classical source or a laser, the probability that n photons be incident on the beamsplitter in one micro-second is given by a Poisson law

$$\mathrm{Prob}(n) = \frac{\mathrm{e}^{-\langle n\rangle}\langle n\rangle^n}{n!}.$$

(11.17)

The probability of observing two photons in one micro-second is $\simeq \langle n\rangle^2/2$ ($\mathrm{e}^{-\langle n\rangle} \simeq 1$ when $\langle n\rangle = 0.1$), and it is then about 5%. In quantum cryptography, one uses typically attenuated laser pulses containing on average 0.1 photons, so that the probability for the pulse to contain two photons is 5%. Physicists can build photon sources where the law (11.17) is not satisfied. This is of course the case for single photon sources where $\mathrm{Prob}(n = 1) = 1$, $\mathrm{Prob}(n \neq 1) = 0$, and more generally for states called Fock states, where the number of photons is fixed. Another example is that of squeezed sates, where the Poisson law is not valid.

A.3.1 Proof of Bell's inequality

Let us start with the example given in Section 3.2 and examine the product $T_A P_B$: this product is equal to $+1$ if $T_A = P_B$ (Alice's T-shirt has the same color as Bob's pants) and -1 if $T_A \neq P_B$ (Alice's T-shirt and Bob's pants have different colors). Alice and Bob perform a large number N of trials, and note for each trial the value of $T_A P_B$, which we denote $[T_A P_B]_n$ for trial number n. The *mean value* of $T_A P_B$, denoted $\langle T_A P_B \rangle$, is the sum of all the results divided by N

$$\langle T_A P_B \rangle = \frac{1}{N}\{[T_A P_B]_1 + \cdots + [T_A P_B]_n + \cdots + [T_A P_B]_N\}$$

$$= \frac{1}{N} \sum_{n=1}^{N} [T_A P_B]_n. \tag{11.18}$$

If the clothes are assigned according to a probability law $\mathrm{Prob}(T_A, T_B, P_A, P_B)$, the mean value of $\langle T_A P_B \rangle$ is by definition

$$\langle T_A P_B \rangle = \sum_{T_A, T_B, P_A, P_B} T_A P_B \, \mathrm{Prob}(T_A, T_B, P_A, P_B).$$

The mean value is expressed as a function of the probabilities for $T_A = T_B$ and $T_A \neq P_B$

$$\langle T_A P_B \rangle = \mathrm{Prob}(T_A = P_B) - \mathrm{Prob}(T_A \neq P_B)$$
$$\langle T_A P_B \rangle = 2\,\mathrm{Prob}(T_A = P_B) - 1 = 1 - 2\,\mathrm{Prob}(T_A \neq P_B), \tag{11.19}$$

where the second line is deduced from the first one by observing that

$$\mathrm{Prob}(T_A = P_B) + \mathrm{Prob}(T_A \neq P_B) = 1.$$

Let us consider the following combination $\langle X \rangle$ of mean values

$$\langle X \rangle = \langle T_A T_B \rangle + \langle T_A P_B \rangle + \langle P_A T_B \rangle - \langle P_A P_B \rangle$$
$$= \langle T_A(T_B + P_B) + P_A(T_B - P_B) \rangle.$$

If $T_B = -P_B$, the first term in the bracket vanishes, and the second one is equal to $2P_A T_B$. Since $P_A = \pm 1$, $T_B = \pm 1$, we obtain $X = \pm 2$. If, on the contrary, $T_B = P_B$, it is now the second term which vanishes, but we find once more $X = \pm 2$: in all cases, $X = \pm 2$. In the calculation of the mean value, it may happen that X be positive or negative, but in all cases the absolute value of the mean value will be less than, or equal to 2: $|\langle X \rangle| \le 2$

$$|\langle T_A T_B \rangle + \langle T_A P_B \rangle + \langle P_A T_B \rangle - \langle P_A P_B \rangle| \le 2. \tag{11.20}$$

This inequality takes a more intuitive form if it is expressed in term of probabilities thanks to (11.19)

$$\text{Prob}(T_A = T_B) + \text{Prob}(T_A = P_B) + \text{Prob}(P_A = T_B)$$

$$+ \text{Prob}(P_A \neq P_B) \le 3, \tag{11.21}$$

where, for example, $\text{Prob}(T_A = P_B)$ is the probability that Alice's T-shirt and Bob's pants have the same color, or $\text{Prob}(P_A \neq P_B)$ the probability that Alice's and Bob's pants have a different color. Inequality (11.20) or, equivalently, (11.21) is an example of Bell's inequality.

We now wish to complete the proof of the factorization property (3.7). Let us first recall Bayes's rule in probability theory. If $\text{Prob}(A, B)$ is a joint probability for two random variables A and B, we define the marginals as

$$\text{Prob}(A) = \sum_B \text{Prob}(A, B) \quad \text{Prob}(B) = \sum_A \text{Prob}(A, B).$$

Bayes's rule states that the *conditional* probability $\text{Prob}(A|B)$, that is, the probability of A given B, is obtained as

$$\text{Prob}(A|B) = \text{Prob}(A, B)/\text{Prob}(B). \tag{11.22}$$

Now comes the crux of the argument: given that the set (X, λ) is assumed to account *completely* for the theoretical description of the experiment, any information other than that contained in the intersection of Σ (see Figure 3.5) with the past light cone of R_A is irrelevant for results in R_A. Let us consider the probability $\text{Prob}(A|a, b, B, X, \lambda)$ for outcome A, conditioned on (a, b, X, λ) *and* the outcome B. Any information on outcome B is redundant and we may write

$$\text{Prob}(A|a, b, B, X, \lambda) = \text{Prob}(A|a, b, X, \lambda) = \text{Prob}(A|a, X, \lambda), \tag{11.23}$$

where we have used the fact that $\text{Prob}(A|a, b, X, \lambda)$ cannot depend on b, because the choice of b cannot influence outcome A. Then, from Bayes's rule (11.22)

$$\text{Prob}(A, B|a, b, X, \lambda) = \text{Prob}(A|B, a, b, X, \lambda)\,\text{Prob}(B|a, b, X, \lambda)$$
$$= \text{Prob}(A|a, X, \lambda)\,\text{Prob}(B|b, X, \lambda).$$

which proves (3.7). In order to derive Bell's inequality, we have only to observe as above that

$$AB + A'B + AB' - A'B' = A(B + B') + A'(B - B') = \pm 2,$$

so that if we integrate this quantity over λ with the weight

$$\text{Prob}(A|a, X, \lambda)\,\text{Prob}(B|b, X, \lambda)\rho(\lambda),$$

then the absolute value of the result, that is, $|\langle AB + A'B + AB' - A'B'\rangle|$, must be ≤ 2.

A.3.2 Entangled states

This appendix develops the mathematical description of entangled states. More precisely, we address the description of two photons in a state entangled in polarization, but the formalism is easily generalized to other cases. We gave in Appendix A.3.2 the mathematical description of the polarization state $|\theta\rangle$ of a photon, where the photon polarization is oriented along a vector making an angle θ with the vertical axis of a plane perpendicular to the direction of propagation (see (11.13))

$$|\theta\rangle = \cos\theta|V\rangle + \sin\theta|H\rangle. \tag{11.24}$$

As we are interested in the two-photon case, we shall distinguish the two photons by indices A or B: for example, $|V_A\rangle$ represents the state of the first photon $\mathcal{A}$ with a vertical polarization, while $|H_B\rangle$ is the state of the second photon $\mathcal{B}$ with a horizontal polarization. In order to describe the photon pair, we need to consider four possibilities:

$$|V_A\rangle|V_B\rangle \quad |V_A\rangle|H_B\rangle \quad |H_A\rangle|V_B\rangle \quad |H_A\rangle|H_B\rangle.$$

For example, $|V_A\rangle|H_B\rangle$ represents a state in which photon $\mathcal{A}$ is vertically polarized, while photon $\mathcal{B}$ is horizontally polarized. We may obtain a more general situation $|\theta_A\rangle|\theta_B\rangle$ by using (11.24) for the two photons

$$|\theta_A\rangle = \cos\theta|V_A\rangle + \sin\theta|H_A\rangle,$$
$$|\theta_B\rangle = \cos\theta|V_B\rangle + \sin\theta|H_B\rangle. \tag{11.25}$$

We could even use two different angles for the two lines of (11.25), but we shall not need it. A state such as $|V_A\rangle|H_B\rangle$ or $|\theta_A\rangle|\theta_B\rangle$ is a called a *tensor product*. In the same manner as we built a linear combination of states $|V\rangle$ and $|H\rangle$ in (11.24), nothing prevents us from building a linear superposition of two-photon states, for example $|\Phi\rangle$

$$|\Phi\rangle = \frac{1}{\sqrt{2}}(|V_A\rangle|V_B\rangle + |H_A\rangle|H_B\rangle). \tag{11.26}$$

This state, which is a linear combination of two tensor products, is an *entangled state*. In this state, there is a 100% correlation between the polarizations for measurements in the $\{VH\}$ basis: if a measurement of photon $\mathcal{A}$ polarization gives V (resp. H), then that of photon $\mathcal{B}$ will also give V (resp. H). However, the results of Alice's measurements are random: she has 50% chance of finding a vertical polarization and 50% of finding a horizontal one. Indeed, (11.26) shows that the probability amplitude of finding $|V_A\rangle$ is $1/\sqrt{2}$, and it is also the probability amplitude of finding $|H_A\rangle$. This entangled state, which was used in the discussion of Chapter 3, possesses a quite remarkable property: its form does not change under rotations around the propagation axis. In order to show this, let us introduce the polarization state $|\theta_\perp\rangle$ orthogonal to $|\theta\rangle$

$$|\theta_\perp\rangle = -\sin\theta|V\rangle + \cos\theta|H\rangle. \tag{11.27}$$

Elementary algebra shows that $|\Phi\rangle$ can also be written as

$$|\Phi\rangle = \frac{1}{\sqrt{2}}(|\theta_A\rangle|\theta_B\rangle + |\theta_{A\perp}\rangle|\theta_{B\perp}\rangle). \tag{11.28}$$

One has only to use (11.25) and (11.27) in (11.28) and $\cos^2\theta + \sin^2\theta = 1$ to recover the initial expression (11.26) of $|\Phi\rangle$. The polarizations are 100% correlated, whatever the orientation used in a plane perpendicular to the direction of propagation.

Let us finally demonstrate equation (3.4). Due to rotational invariance, we may choose $\vec{a}$ along the vertical axis and $\vec{b}$ making an angle θ with this direction. Let us assume that we measure a vertical polarization for photon $\mathcal{A}$. Then, from (11.26), the polarization of photon $\mathcal{B}$ is also vertical, and the state of this photon is $|V_B\rangle$. We now invert (11.27)

$$|V_B\rangle = \cos\theta|\theta_B\rangle - \sin\theta|\theta_{B\perp}\rangle.$$

The probability amplitude of finding $|V_B\rangle$ in the polarization state $|\theta_B\rangle$, that is, a polarization oriented along $\vec{b}$, is then $\cos\theta$, and the corresponding probability $\cos^2\theta$, which is nothing other than Malus's law. The joint probability that photon $\mathcal{A}$ has a polarization along $\vec{a}$ and photon $\mathcal{B}$ along $\vec{b}$ is then given by

$$\frac{1}{2}\cos^2\theta = \frac{1}{4}(1 + \cos 2\theta).$$

Because of rotational invariance, the probability that the polarizations be perpendicular to $\vec{a}$ and $\vec{b}$ is given by the same formula, which completes the proof of equation (3.4).

The use of entanglement allows us to shed light on the so-called wave–particle duality, thanks to the following experiment. A source produces pairs of entangled photons in the state (11.26), which we recall for convenience

$$|\Phi\rangle = \frac{1}{\sqrt{2}}(|H\rangle|H\rangle + |V\rangle|V\rangle).$$

The right hand photon, the test photon, is sent to a Mach–Zehnder interferometer and the left hand one, the companion photon, toward a PBS via an optical fiber, so that this second photon is delayed by 20 ns with respect to the test photon. This experiment is called delayed choice because the companion photon is detected 20 ns *after* the test one has already been registered. Furthermore, the polarization of the companion photon may be rotated by an angle α before being detected. The Mach–Zehnder interferometer is of the type sketched on Figure 1.6, with a variable phase shift δ introduced on the blue path. However, the second beamsplitter BS_2 is of a special kind: it reflects 100% of the horizontally polarized photons, but act as an ordinary 50/50 balanced beamsplitter for the vertically polarized photons: BS_2 is a polarization dependent beamsplitter. The net outcome is that the interferometer is open for horizontally polarized photons, which behave as "particles" with no interference, and closed for vertically polarized photons, which behave as "waves" with interference. The two-photon state vector, just after the test photon has left the interferometer, is

$$|\Phi\rangle = \frac{1}{2}|H_C\rangle(|H,X\rangle + e^{i\delta}|H,Y\rangle)$$

$$+ \frac{1}{2\sqrt{2}}|V_C\rangle([e^{i\delta} + 1]|V,X\rangle + [e^{i\delta} - 1]|V,Y\rangle). \quad (11.29)$$

The notation is as follows: $|H_C\rangle$ ($|V_C\rangle$) denotes the state of the companion photon polarized horizontally (vertically) and $|H,X\rangle$ a horizontally polarized test photon leaving the beamsplitter BS_2 along the OX direction. Equation (11.29) may be rewritten as

$$|\Phi\rangle = \frac{1}{2}|H_C\rangle|P\rangle + \frac{1}{2}|V_C\rangle|W\rangle, \quad (11.30)$$

where

$$|P\rangle = |H,X\rangle + e^{i\delta}|H,Y\rangle,$$

$$|W\rangle = \frac{1}{\sqrt{2}}([e^{i\delta} + 1]|V,X\rangle + [e^{i\delta} - 1]|V,Y\rangle).$$

In the preceding equation, $|P\rangle$ denotes a particle-like behavior and $|W\rangle$ a wave-like one. Now, we rotate the polarization of the companion photon by an angle α before its detection. If the state vector of the companion photon is

$$\cos\alpha|H_C\rangle - \sin\alpha|V_C\rangle,$$

then that of the test photon is

$$|\Psi\rangle = \cos\alpha(e^{i\delta}|H,X\rangle + |H,Y\rangle)$$

$$- \frac{1}{\sqrt{2}}\sin\alpha([e^{i\delta} + 1]|V,X\rangle + [e^{i\delta} - 1]|V,Y\rangle). \quad (11.31)$$

In terms of probability amplitudes, it means that (see Section 9.1 for the notations)

$$a_X(H) = \cos\alpha, \qquad a_X(V) = -\frac{1}{\sqrt{2}}\sin\alpha[e^{i\delta} + 1]$$

$$a_Y(H) = e^{i\delta}\cos\alpha, \quad a_Y(V) = -\frac{1}{\sqrt{2}}\sin\alpha[e^{i\delta} - 1].$$

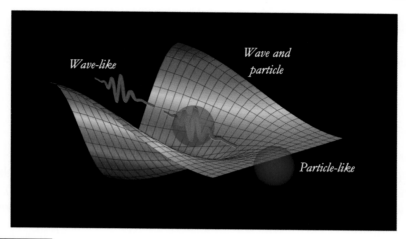

Wave-like

Wave and particle

Particle-like

Figure 11.6. Artist's view of photon behavior. One observes a continuous transition from a wave behavior (background of the drawing) to a particle one (front part of the drawing). Adapted from Kaiser *et al.* [2012]. Copyright: F. Kaiser and S. Tanzilli. Courtesy of Sébastien Tanzilli.

If the test photons are registered by a polarization insensitive detector after BS_2, we must add the probabilities for horizontally polarized and vertically polarized photons (rule number 3 in § 1.6.2) so that, for example, in the horizontal direction OX

$$\text{Prob}_X(\alpha, \delta) = |a_X(H)|^2 + |a_X(V)|^2$$

$$= \cos^2 \alpha + \frac{1}{2} \sin^2 \alpha |e^{i\delta} + 1|^2 = 1 + \sin^2 \alpha \cos \delta. \quad (11.32)$$

Figure 11.6 displays the result (11.32) as a function of the two angles α and δ. We observe a continuous transition from the "particle" aspect for $\alpha = 0$ to the "wave" aspect for $\alpha = \pi/2$. The two limit cases are particle and wave: in the particle case (front part of the figure), the probability is independent of δ, and in the wave case (back part of the figure), it is given by the standard $(1 + \cos \delta)$ law (1.3) of interferences. The configuration of the experiment allows one to check the violation of Bell's inequalities, so that the two-photon quantum state is indeed an entangled state, and not an incoherent mixture. The conclusion is that we are able to observe the wave and particle behaviors in a single experiment, in contradiction to the usual statement of wave–particle duality found in most textbooks.

A.4.1 The formalism of quantum theory

It is convenient divide our summary of quantum theory into three stages.

1. The mathematical description of a quantum state.
2. The probabilities and Born's rule.
3. The time evolution of a quantum state and Schrödinger's equation.

Let us begin with the description of a quantum state. We gave in Appendix A.3.2 the description of the polarization quantum state of a photon. This quantum state is represented mathematically by a vector $|\varphi\rangle$ of a two-dimensional space, and we may choose in this space two basis vectors $|V\rangle$ and $|H\rangle$ describing vertical and horizontal polarization states. These two states are orthogonal: the scalar product $\langle H|V\rangle = 0$, and they are of length, or more correctly of *norm*, unity. The norm $\|\varphi\|$ of a vector $|\varphi\rangle$ generalizes the notion of length: it is the square root of the scalar product $\langle\varphi|\varphi\rangle$, $\|\varphi\| = \sqrt{\langle\varphi|\varphi\rangle}$ and we have $\langle V|V\rangle = \langle H|H\rangle = 1$. The most general polarization state is described by a vector $|\varphi\rangle$, the (polarization) *state vector*, which is a linear combination with complex coefficients c_V and c_H of $|V\rangle$ and $|H\rangle$ of type (11.15)

$$|\varphi\rangle = c_V|V\rangle + c_H|H\rangle, \quad |c_H|^2 + |c_V|^2 = 1. \qquad (11.33)$$

The condition $|c_H|^2 + |c_V|^2 = 1$ ensures that the vector $|\varphi\rangle$ has norm unity: $\langle\varphi|\varphi\rangle = 1$.

In the general case, a quantum state is described mathematically by a vector $|\varphi\rangle$ which is an element of a vector space $\mathcal{H}$, the *space of states*, which we choose for simplicity to be of finite dimension D. Infinite dimensional spaces are compulsory in some circumstances, but they lead to serious mathematical complications which we prefer to avoid. The case of photon polarization corresponds to $D = 2$. A D-dimensional space is spanned by D orthonormal basis vectors $|1\rangle, \ldots, |n\rangle, \ldots, |D\rangle$, $\langle i|j\rangle = \delta_{ij}$

$$|\varphi\rangle = c_1|1\rangle + \cdots + c_n|n\rangle + \cdots + c_D|D\rangle, \qquad (11.34)$$

with the normalisation condition

$$\langle\varphi|\varphi\rangle = |c_1|^2 + \cdots + |c_n|^2 + \cdots + |c_D|^2 = 1.$$

If we consider another vector $|\chi\rangle$

$$|\chi\rangle = b_1|1\rangle + \cdots + b_n|n\rangle + \cdots + b_D|D\rangle, \qquad (11.35)$$

then the scalar product $\langle\chi|\varphi\rangle$ is

$$\langle\chi|\varphi\rangle = b_1^* c_1 + \cdots + b_n^* c_n + \cdots + b_D^* c_D. \qquad (11.36)$$

In the second part of this Section, we give a mathematical expression to the probability amplitudes introduced in § 1.6.2, rule 2. If the state vectors $|\varphi\rangle$ and $|\chi\rangle$ are given by (11.34) and (11.35), the *probability amplitude* of finding $|\varphi\rangle$ in $|\chi\rangle$ is the scalar product $\langle\chi|\varphi\rangle$ (11.36) and the corresponding probability is $|\langle\chi|\varphi\rangle|^2$. These two definitions are rather abstract, but they will take a more intuitive form thanks to *Born's rule* which we first illustrate on the photon polarization case. Let us assume that the polarization state vector is given by (11.15)

$$|\varphi\rangle = \cos\theta|V\rangle + e^{i\delta}\sin\theta|H\rangle,$$

and let this photon go through a polarizing beamsplitter (PBS) which separates photons with vertical polarization from those with horizontal polarization (Figure 2.2). The probability Prob_V of triggering the detector D_V is given by the modulus squared of the scalar product $\langle V|\varphi\rangle$, which is the probability amplitude of finding $|\varphi\rangle$ in $|V\rangle$

$$\text{Prob}_V = |\langle V|\varphi\rangle|^2 = \cos^2\theta.$$

Similarly, the probability Prob_H of triggering the detector D_H is given by

$$\text{Prob}_H = |\langle H|\varphi\rangle|^2 = \sin^2\theta.$$

In the general case, if the quantum state is described by a vector in a D-dimensional space, a measurement will involve a "multichannel PBS" with D output channels $|1\rangle, \cdots, |n\rangle, \cdots, |D\rangle$ (Figure 11.7). The probability that the quantum system chooses channel $|n\rangle$ is

$$|\langle n|\varphi\rangle|^2 = |c_n|^2. \qquad (11.37)$$

When $|\varphi\rangle$ is replaced by $e^{i\delta}|\varphi\rangle$, the probabilities $|\langle\chi|\varphi\rangle|^2$ are unchanged. As experiment gives access to probabilities only, the vectors $|\varphi\rangle$ and $e^{i\delta}|\varphi\rangle$ describe the same physical state. Thus (11.15) describes the most general

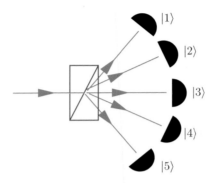

$|1\rangle$

$|2\rangle$

$|3\rangle$

$|4\rangle$

$|5\rangle$

Figure 11.7.   Multichannel polarizing beamsplitter, with $D = 5$ output channels.

polarization state, as it is always possible to take the coefficient c_V of $|V\rangle$ in (11.27) as a real number: $c_V = \cos\theta$.

One important notion is that of *incompatible bases*, or *complementary bases*, which we explain using polarization. Instead of using the basis $\{|V\rangle, |H\rangle\}$, we may use that defined by the states $|D\rangle$ and $|A\rangle$ (11.14). A measurement in this new basis will be performed by rotating the PBS in Figure 2.2 by $45°$ around the propagation axis of the incident photons. The measurement is then performed in the $\{|D\rangle, |A\rangle\}$ basis, instead of the $\{|V\rangle, |H\rangle\}$ one. Let us assume that an incident photon is polarized in the vertical direction. If the measurement is made in the $\{|V\rangle, |H\rangle\}$ basis, the polarization will be determined to be vertical with a 100% probability: the measurement result is certain. However, if the measurement is made in the $\{|D\rangle, |A\rangle\}$ basis, we shall find a $+45°$, or D, polarization with a 50% probability and a $-45°$, or A, polarization with a 50% probability. While the value of the polarization of the incident photon is well-defined in the $\{|V\rangle, |H\rangle\}$ basis, that is not the case in the $\{|D\rangle, |A\rangle\}$ one: the two bases are termed incompatible. This is exactly the situation for position and momentum measurements, x and p: these two variables are incompatible. If the value of x is certain, that of p is completely undetermined, and vice-versa.

In a third step, we tackle time evolution. Up to now, we have given the description of a quantum state at a fixed instant of time, for example, at a time $t = t_0$, and described the measurement's operation on this state. A complete description requires that we are able to answer the question: what is the time evolution of this state for times $t > t_0$? This evolution is unitary, it involves an operator $U(t, t_0)$: a unitary operator U is such that the scalar product is conserved, $\langle U\varphi | U\chi \rangle = \langle \varphi | \chi \rangle$, whatever the vectors $|\varphi\rangle$ and $|\chi\rangle$. The state vector at time t is given as a function of the state vector $|\varphi(t_0)\rangle$ at $t = t_0$ by

$$|\varphi(t)\rangle = U(t, t_0)|\varphi(t_0)\rangle. \tag{11.38}$$

When the quantum system is isolated, the choice of an origin of time is arbitrary, and we must have whatever the time interval τ

$$U(t + \tau, t_0 + \tau) = U(t, t_0).$$

This result implies that $U(t, t_0)$ can depend only on the difference $(t - t_0)$. A theorem due to Stone then shows that $U(t - t_0)$ takes the form

$$U(t - t_0) = e^{-iH(t-t_0)/\hbar}. \tag{11.39}$$

Since the argument in the exponential must be dimensionless, H has the dimension of an energy and, in fact, H is the energy operator, or *Hamiltonian*, of the quantum system. Combining (11.38) and (11.39) we deduce

$$|\varphi(t)\rangle = e^{-iH(t-t_0)/\hbar}|\varphi(t_0)\rangle,$$

and if we take the derivative with respect to t

$$\frac{d}{dt}|\varphi(t)\rangle = -\frac{iH}{\hbar}e^{-iH(t-t_0)/\hbar}|\varphi(t_0)\rangle = -\frac{iH}{\hbar}|\varphi(t)\rangle,$$

which is usually written

$$\boxed{i\hbar\frac{d}{dt}|\varphi(t)\rangle = H|\varphi(t)\rangle}. \tag{11.40}$$

This is the *evolution equation*, better known as the *Schrödinger equation*. It remains valid even if H depends on time, $H \to H(t)$. The remarkable feature of equations (11.38) or (11.40) is that they are *deterministic* equations: given the initial conditions at time $t = t_0$ in the form of a state vector $|\varphi(t_0)\rangle$, they determine the state vector at any later time t, exactly as the equations of classical mechanics determine the future evolution given the initial positions and velocities. *The probabilistic aspect is introduced only at the time of measurement, through Born's rule*, but between two measurements, the evolution is fully deterministic.

When the Hamiltonian is time-independent, we can show that there exist D vectors, not necessarily unique, $|\varphi_1\rangle, \ldots, |\varphi_n\rangle, \ldots, |\varphi_D\rangle$, which form an orthonormal basis of the space of states $\mathcal{H}$, with

$$H|\varphi_n\rangle = E_n|\varphi_n\rangle. \tag{11.41}$$

The states $|\varphi_n\rangle$ are called *stationary states*, because their time evolution is very simple: if $|\varphi(t_0)\rangle = |\varphi_n\rangle$, then

$$|\varphi(t)\rangle = e^{-iH(t-t_0)/\hbar} |\varphi(t_0)\rangle = e^{-iE_n(t-t_0)/\hbar} |\varphi_n\rangle$$
$$= e^{-iE_n(t-t_0)/\hbar} |\varphi(t_0)\rangle, \tag{11.42}$$

because the action of H on $|\varphi_n\rangle$ is simply $H \rightarrow E_n$. Time evolution corresponds then to a mere multiplication by a phase factor and, as we have seen, this does not change the physical state: a state such as (11.41) stays physically unchanged when time varies, hence the terminology.

A.4.2 A quantum particle in a one-dimensional box

Let us return to the example of § 4.1, that of a particle in a one-dimensional box, and let $|x\rangle$ be a state where the particle has a well-defined position: it is located precisely at point x. Strictly speaking, this is forbidden by Heisenberg's inequality, because it would imply an infinite momentum, and therefore an infinite energy. Quantum states such as $|x\rangle$ may be used, but with some caution. The difficulty is linked to the fact that we should use an infinite dimensional space of states and face the corresponding mathematical problems. If the quantum state of the particle is $|\psi\rangle$, then the probability amplitude of finding it at point x is $\langle x|\psi\rangle = \psi(x)$, the particle wave function. Let $|\psi_n\rangle$ be a state of the form (11.41)

$$H|\psi_n\rangle = E_n|\psi_n\rangle$$

and multiply this equation on the left by $\langle x|$

$$\langle x|H|\psi_n\rangle = E_n\langle x|\psi_n\rangle = E_n\psi_n(x).$$

One can show that

$$\langle x|H|\psi\rangle = -\frac{\hbar^2}{2m^2} \frac{d^2\psi(x)}{dx^2}.$$

The proof is somewhat technical and can be found in any quantum mechanics textbook. This gives the time-independent Schrödinger equation for $\psi_n(x)$

$$-\frac{\hbar^2}{2m^2}\frac{d^2\psi_n(x)}{dx^2} = E_n\psi_n(x), \tag{11.43}$$

or

$$\frac{d^2\psi_n(x)}{dx^2} + k_n^2\psi_n(x) = 0, \tag{11.44}$$

with $k_n^2 = 2mE_n/\hbar^2$. The solutions of (11.44) are linear combinations of $\sin k_n x$ and $\cos k_n x$, but since $\psi_n(x)$ must vanish at $x = 0$, we keep only the solution $\sin k_n x$. Morover, $\psi_n(x)$ must vanish at $x = L$, which implies $k_n L = n\pi$, $n = 1, 2 \ldots$ One finds the results quoted in § 4.2, $k_n = n\pi/L$ and energy level quantization in the form of $E_n = \hbar^2 n^2 \pi^2/(2mL^2)$.

The quantity $|\psi_n(x)|^2\delta x$ is the probability of finding the particle in the interval $[x, x + \delta x]$, for small enough δx, in the present case $\delta x \ll L$. As the particle must be found somewhere in the $[0, L]$ interval

$$1 = \int_0^L |\psi_n(x)|^2 dx = a_n^2 \int_0^L \sin^2\left(\frac{\pi n x}{L}\right) dx = a_n^2 \frac{L}{2} \tag{11.45}$$

hence $a_n = \sqrt{2/L}$.

To conclude this Appendix, let us use Heisenberg's inequality (4.3) to estimate the energy of the one-dimensional harmonic oscillator ground state and that of the hydrogen atom. The Hamiltonian of the harmonic oscillator, for example, that of the trap in § 5.3.2 is of the form kinetic+potential energy

$$H = \frac{p^2}{2m} + \frac{1}{2}m\Omega^2 x^2,$$

where $\Omega = 2\pi f$ is the angular frequency of the trap. Using Heisenberg's inequality in the form $p = \hbar/2x$, we write H as

$$H = \frac{\hbar^2}{8m^2x^2} + \frac{1}{2}m\Omega^2 x^2.$$

We look for the minimum of H: $dH/dx = 0$, which leads to $x^2 = \hbar/(2m\Omega)$, and plugging this result into H

$$H = \frac{1}{4}\hbar\Omega + \frac{1}{4}\hbar\Omega = \frac{1}{2}\hbar\Omega. \tag{11.46}$$

This coincides with the exact result. As a matter of fact, the ground state wave function of the harmonic oscillator is said to "saturate" Heisenberg's inequality, meaning that the inequality becomes an equality. This is the reason why we get the exact result, because writing $\Delta p \Delta x = \hbar/2$ is exact for this particular wave function.

We now turn to the hydrogen atom. If the electron traces a circular orbit of radius r with a momentum $p = mv$, its classical energy is

$$E = \frac{p^2}{2m} - \frac{e^2}{r}. \tag{11.47}$$

In classical physics, the orbit radius tends to zero, "the electron falls into the nucleus", this phenomenon being accompanied by emission of electromagnetic radiation. Indeed, in classical physics, nothing prevents the orbit radius from becoming arbitrarily small. The decrease in the orbit radius is accompanied by a decrease in the orbit energy, but this is compensated by electromagnetic energy emission into space, which ensures energy conservation. The situation is changed radically in quantum physics, due to Heisenberg's inequality. For an orbit of radius r, the dispersion Δx of the position along an axis is on the order of r, so that the momentum dispersion is at least on the order of $\sim \hbar/\Delta x = \hbar/r$. We infer $rp \sim \hbar$ and the expression (11.47) of the energy becomes

$$E \sim \frac{\hbar^2}{2mr^2} - \frac{e^2}{r}.$$

Let us look for the minimum of E

$$\frac{dE}{dr} \sim -\frac{\hbar^2}{mr^3} + \frac{e^2}{r^2} = 0,$$

which yields a minimum for

$$r = a_0 = \frac{\hbar^2}{me^2}. \tag{11.48}$$

This nothing other than Bohr's radius of the hydrogen atom. This allows us to recover the ground state energy

$$E_0 = -\frac{e^2}{2a_0} = -\frac{me^4}{2\hbar^2}.$$ (11.49)

This is the exact result for the energy! Contrary to the case of the harmonic oscillator, this coincidence is a numerical accident. However, the underlying reasoning does give a fundamental explanation for the stability of the hydrogen atom, and in fact of all atoms. Owing to Heisenberg's inequality, the electron cannot trace an orbit with a too small radius, because otherwise, it would acquire a huge kinetic energy. Stability is obtained by looking for the best compromise between kinetic and potential energy, so that the ground state energy is minimal.

A.10.1 The de Broglie–Bohm theory

Let us write the wave function $\psi(\vec{r}, t)$ in a modulus/phase representation ($\hbar = h/2\pi$)

$$\psi(\vec{r}, t) = R(\vec{r}, t) \exp(iS(\vec{r}, t)/\hbar).$$

From the Schrödinger equation, we deduce the time evolution of the phase S

$$\frac{\partial S}{\partial t} + \frac{1}{2m} (\vec{\nabla}S)^2 + V + U = 0,$$

where $\vec{\nabla}$ is a gradient, V the ordinary potential and U the quantum potential, $U = -(\hbar^2/2m)(\nabla^2 R)$. If $\hbar = 0$, we recover the equations of classical mechanics in the Hamilton–Jacobi form. The momentum $\vec{p}$ is the gradient of the phase: $\vec{p} = \vec{\nabla}S$. The modified Newton's law is

$$\frac{d\vec{p}}{dt} = \frac{d(\vec{\nabla}S)}{dt} = -\vec{\nabla}(V + U),$$

where d/dt is calculated along the trajectory. The wave function obeys the Schrödinger equation (11.43) and is not influenced by the position of the particle.

References

1. Adler [2006]: R. Adler, Gravity, in *The new physics for the twentieth century*, p. 41, Cambridge University Press, Cambridge.
2. Arndt *et al.* [2005]: M. Arndt, K. Hornberger and A. Zeilinger, Probing the limits of the quantum world, *Physics World* **18**, March 2005, pp. 35–40.
3. Aspect [1999]: A. Aspect, Bell's inequalities: more ideal than ever, *Nature* **398**, 189–190.
4. Aspect *et al.* [1982]: A. Aspect, P. Grangier and G. Roger, Experimental realization of Einstein-Podolsky-Rosen gedanken experiment: a new violation of Bell's inequalities, *Phys. Rev. Lett.* **49**, 91; A. Aspect, J. Dalibard and G. Roger, Experimental test of Bell's inequalities using time varying analyzers, *Phys. Rev. Lett.* **49** 1804.
5. Aspect *et al.* [1989]: A. Aspect, P. Grangier and G. Roger, Dualité onde-corpuscule pour un photon unique, *J. Optics (Paris)* **20**, 119.
6. Ballentine [1998]: L. Ballentine, *Quantum Mechanics*, World Scientific.
7. Bell [2004]: J. S. Bell, *Speakable and unspeakable in quantum mechanics*, Cambridge University Press, Cambridge.
8. Blatt [2004]: R. Blatt, Quantum information processing in ion traps, in Les Houches Summer School 2003, *Quantum entanglement and information processing*, D. Estève, J.-M. Raimond, and M. Brune eds., Elsevier, Amsterdam, pp. 223–260.
9. Bloch [2013]: I. Bloch, Quantum leaps for simulations, *Physics World* **26**, March 2013, pp. 47–51.
10. Bohr [1935]: N. Bohr, Can quantum mechanical description of physical reality be considered complete?, *Phys. Rev.* **48**, 696.
11. Brune *et al.* [1996b]: M. Brune, E. Hagley, J. Dreyer, X. Maître, A. Maali, C. Wunderlich, J-M Raimond and S. Haroche, Observing the progressive decoherence of the "meter" in a quantum measurement, *Phys. Rev. Lett.* **77**, 4887.

12. Clauser [1976]: J. Clauser, Experimental investigation of a polarization correlation anomaly, *Phys. Rev. Lett.* **36**, 1223.

13. Cohen-Tannoudji and Dalibard [2006]: C. Cohen-Tannoudji and J. Dalibard, Manipulating atoms with photons, in *The new physics for the twentieth century*, p. 145, Cambridge University Press, Cambridge.

14. Cohen-Tannoudji *et al.* [1977]: C. Cohen-Tannoudji, B. Diu and F. Laloë, *Quantum mechanics*, John Wiley, New York.

15. Einstein *et al.* [1935]: A. Einstein, B. Podolsky and N. Rosen, Can quantum-mechanical description of physical reality be considered complete?, *Phys. Rev.* **47**, 777–780.

16. Ekert [2006]: A. Ekert, Quanta, ciphers and computers, in *The new physics for the twentieth century*, p. 268, Cambridge University Press, Cambridge.

17. d'Espagnat [2003]: B. d'Espagnat, *Veiled reality*, Westview Press Inc.

18. Feynman *et al.* [1965]: R. Feynman, R. Leighton and M. Sands, *The Feynman Lectures on Physics*, Addison-Wesley, Reading.

19. Freedman and Clauser [1972]: S. Freedman and J. Clauser, Experimental test of local hidden variable theories, *Phys. Rev. Lett.*, **28**, 938.

20. Gisin [2013]: N. Gisin, Quantum correlations in Newtonian space and time: arbitrarily fast communication and non-locality, arxiv: 1210.7308.

21. Gisin *et al.* [2002]: N. Gisin, G. Ribordy, W. Tittel and H. Zbinden, Quantum cryptography, *Rev. Mod. Phys.* **74**, 145–195.

22. Gleyzes *et al.* [2007]: S. Gleyzes *et al.*, Quantum jumps of light recording the birth and death of a photon in a cavity, *Nature* **446**, 297.

23. Green [1999]: B. Green, *The elegant universe: superstrings, hidden dimensions and the quest for the ultimate theory*, Vintage Series, Random House Inc.

24. Greene [2006]: M. Greene, Superstring theory, in *The new physics for the twentieth century*, p. 41, Cambridge University Press, Cambridge.

25. Grynberg *et al.* [2010]: G. Grynberg, A. Aspect and C. Fabre, *Introduction to quantum optics*, Cambridge University Press, Cambridge.

26. Gubser [2010]: S. Gubser, *The little book of string theory*, Princeton University Press, Princeton.

27. Guerlin *et al.* [2007]: C. Guerlin *et al.*, Progressive field-state collapse and quantum non-demolition photon counting, *Nature* **448**, 889.

28. Haroche and Raimond [2006]: S. Haroche and J.-M. Raimond, *Exploring the quantum*, Oxford University Press, Oxford.

29. Hey and Walters [2003]: T. Hey and P. Walters, *The new quantum universe*, Cambridge University Press, Cambridge.

30. Holland [1993]: P. Holland, *The quantum theory of motion*, Cambridge University Press, Cambridge.

31. Howard [2004]: D. Howard, Who invented the Copenhagen interpretation?, *Phil. of Science* **71**, 669.

32. Jacques *et al.* [2005]: V. Jacques, E. Wu, F. Grosshans, F. Treussart, Ph. Grangier, A. Aspect and J-F Roch, Single photon wave-front splitting interference, *Eur. J. Phys.* **D 35**, 561.

33. Jacques *et al.* [2007]: V. Jacques, E. Wu, F. Grosshans, F. Treussart, Ph. Grangier, A. Aspect and J-F Roch, Experimental realization of Wheeler's delayed choice experiment, *Science* **315**, 966.

34. Kaiser *et al.* [2012]: F. Kaiser, T. Coudreau, P. Milman, D. Ostrowsky and S. Tanzilli, Entanglement enabled delayed choice experiment, *Science* **338**, 637.

35. Landau and Lifschitz [1977]: L. Landau and E. Lifschitz, *Quantum mechanics: non-relativistic theory*, Pergamon Press.

36. Laloë [2012]: F. Laloë, *Do we really understand quantum mechanics?*, Cambridge University Press, Cambridge.

37. Le Bellac [2006]: M. Le Bellac, *A short introduction to quantum information and computation*, Cambridge University Press, Cambridge.

38. Lees *et al.* [2012]: J.-P. Lees *et al.* (Babar collaboration), Observation of time-reversal violation in the B^0-meson system, arxiv: 1207.5832.

39. Leggett [2005]: A. Leggett, The quantum measurement problem, *Science* **307**, 871.

40. Loepp and Wootters [2006]: S. Loepp and W. Wootters, *Protecting information*, Cambridge University Press, Cambridge.

41. Matsukevitch *et al.* [2008]: D. Matsukevitch, P. Maunz, D. Moering, S. Olmscheck and C. Monroe, Bell inequality violation with two remote atomic qubits, *Phys. Rev. Lett.* **100**, 150404.

42. Maudlin [2011]: T. Maudlin, *Quantum non-locality and relativity*, Wiley-Blackwell, John Wiley and Sons.

43. Mermin [2007]: N. Mermin, *Quantum computer science*, Cambridge University Press, Cambridge.

44. Mohiden and Roy [1998]: U. Mohiden and A. Roy, Precision measurements of the Casimir force from 0.1 to 0.9 μm, *Phys. Rev. Lett.* **81**, 4549.

45. Norsen [2011]: T. Norsen, John S. Bell's concept of causal locality, *Am. J. Phys.* **79**, 1261.

46. Paul [2004]: H. Paul, *Introduction to quantum optics*, Cambridge University Press, Cambridge.

47. Peres [1993]: A. Peres, *Quantum Theory, Concepts and Methods*, Kluwer, Boston.

48. Pethick and Smith [2001]: C. Pethick and H. Smith *Bose–Einstein condensates of dilute gases*, Cambridge University Press, Cambridge.

49. Phillips and Foot [2006]: W. Phillips and C. Foot, The quantum world of ultra-cold atoms in *The new physics for the twentieth century*, p. 171, Cambridge University Press, Cambridge.

50. Quigg [2006]: C. Quigg, Particles and the standard model in *The new physics for the twentieth century*, p. 86, Cambridge University Press, Cambridge.

51. Randall [2011]: L. Randall, *Knocking on heaven's door*, Harper & Collins, New York.

52. Rieffel and Polak [2011]: E. Rieffel and W. Polak, *Quantum computing: a gentle introduction*, MIT Press, Cambridge, Ma.

53. Rosencher and Vinter [2002]: E. Rosencher and B. Vinter, *Optoelectronics*, Cambridge University Press, Cambridge.

54. Scarani [2006]: V. Scarani, *Quantum physics: a first encounter, interference, entanglement and reality*, Oxford University Press, Oxford.

55. Scarani [2011]: V. Scarani, Quantum information: primitive notions and quantum correlations, in *Ultracold gases and quantum information*, p. 105, Oxford University Press, Oxford.

56. Scarani *et al.* [2005]: V. Scarani, S. Iblisdir, N. Gisin and A. Acin, Quantum cloning, *Rev. Mod. Phys.* **77**, 1225.

57. Scarani *et al.* [2010]: V. Scarani, Chua Lynn and Liu Shi Yang, *Six quantum pieces*, World Scientific, Singapore.

58. Scheidl *et al.* [2010]: T. Scheidl *et al.*, Violation of local realism with freedom of choice, *Proc. Nat. Sci. USA* **107**, 46.

59. Schlossauer [2007]: M. Schlossauer, *Decoherence and the quantum-to-classical transition*, Springer, Berlin/Heidelberg.

60. Schwarzschild [2012]: B. Schwarzschild, Time-reversal asymmetry in particle physics has finally been clearly seen, *Physics Today* **65** (11), 16.

61. Shimizu *et al.* [1992]: F. Shimizu, K. Shimizu and H. Takuma, Double-slit interference with ultracold metastable neon atoms, *Phys. Rev.* **A46**, R17 (1992).

62. Singh S. [2000]: S. Singh, *The code book: the secret history of codes and code breaking*, Fourth Estate Ltd.

63. Taylor et Wheeler [1963]: E. Taylor and J. Wheeler, *Space-Time Physics*, W. H. Freeman, New York.

64. Smolin [2006]: L. Smolin, *The trouble with physics*, Houghton MifflinCompany, New York.

65. Wark [2005]: D. Wark, Neutrinos: ghosts of matter, *Physics World* **18**(6), June.

66. Weihs *et al.* [1998]: G. Weihs *et al.*, Violation of Bell's inequalities under strict locality conditions, *Phys. Rev. Lett.* **81**, 5039.

67. Zeilinger [2006]: A. Zeilinger, Essential quantum entanglement, in *The new physics for the twentieth century*, p. 257, Cambridge University Press, Cambridge.

68. Zeilinger [2010]: A. Zeilinger, *Dance of the photons*, Farrar, Strauss and Giroux, New York.

69. Zurek [1991]: W. Zurek, Decoherence and the transition from quantum to classical, *Physics Today*, October 1991, p. 36. An updated version (2003) is found in: quant-ph/0306072.

Index